IntechOpen

Supporting open minds since 2005

Nanopores
http://dx.doi.org/10.5772/intechopen.91122
Edited by Sadia Ameen, M. Shaheer Akhtar and Hyung-Shik Shin

Contributors
Mahmoud Fathy Mubarak, Sahar Saad, Per A. Löthman, Sarawati Prasad Mishra, Shweta Dutta, Anil Kumar Kumar Sahu, Koushlesh Mishra, Pankaj Kashyap, Phumlani Tetyana, Poslet Morgan Morgan Shumbula, Zikhona Njengele-Tetyana, Jegatha Nambi Krishnan, Swastic, Ana Rovisco, Sofia Henriques Ferreira, Pedro Barquinha, Rui Igreja, Rodrigo Martins, Hugo Águas, Elvira Fortunato, Andreia Dos Santos, Ahmet Yanik, Xiangchao Zhu, Ahmet Cicek, Yixiang Li, Lekan Taofeek Popoola, Alhaji Shehu Grema, Sadia Ameen

First published in London, United Kingdom, 2021 by IntechOpen
IntechOpen is the global imprint of INTECHOPEN LIMITED, registered in England and Wales, registration number: 11086078, 5 Princes Gate Court, London, SW7 2QJ, United Kingdom
Printed in Croatia

British Library Cataloguing-in-Publication Data
A catalogue record for this book is available from the British Library

Additional hard and PDF copies can be obtained from orders@intechopen.com

Nanopores
Edited by Sadia Ameen, M. Shaheer Akhtar and Hyung-Shik Shin
p. cm.
Print ISBN 978-1-83880-209-7
Online ISBN 978-1-83880-210-3
eBook (PDF) ISBN 978-1-83880-966-9

Meet the editors

Professor Sadia Ameen obtained a Ph.D. in Chemistry in 2008. Presently she is working as an assistant professor in the Department of Bio-Convergence Science, Jeongeup Campus, Jeonbuk National University, Republic of Korea. Her current research focuses on dye-sensitized solar cells, perovskite solar cells, organic solar cells, sensors, catalysts, and optoelectronic devices. She specializes in innovative energy materials and the manufacture of nanocomposites. She received a gold medal in academics and a merit scholarship for her outstanding academic achievements. She is also a recipient of the Best Researcher Award. She has authored or co-authored more than 120 peer-reviewed papers in the fields of solar cells, catalysts, and sensors as well as book chapters and edited books.

Professor M. Shaheer Akhtar obtained a Ph.D. in Chemical Engineering, from Jeonbuk National University, Republic of Korea, in 2008. Presently, he is an associate professor at the same university. His research interests include photo-electrochemical characterizations of thin-film semiconductor nanomaterials, composite materials, polymer-based solid-state films, solid polymer electrolytes, and electrode materials for dye-sensitized solar cells (DSSCs), hybrid organic-inorganic solar cells, small molecule-based organic solar cells, and photocatalytic reactions.

Professor Hyung-Shik Shin received Ph.D. in the kinetics of initial oxidation Al (111) surface from Cornell University, USA, in 1984. He is an Emeritus Professor in the School of Chemical Engineering, Jeonbuk National University and also the President of Korea Basic Science Institute (KBSI), Gwahak-ro, Yuseong-gu, Daejon, Republic of Korea. He has been a promising researcher and visited several universities as visiting professor/invited speaker worldwide. He is an active executive member of various renowned scientific committees such as KiChE, copyright protection, KAERI, etc. He has extensive experience in electrochemistry, renewable energy sources, solar cells, organic solar cells, charge transport properties of organic semiconductors, inorganic-organic solar cells, biosensors, chemical sensors, nano-patterning of thin-film materials, and photocatalytic degradation.

Contents

Preface

The field of nanoporous materials has advanced significantly over the last two decades. Nanoporous materials have a porous structure and are made up of a typical organic or inorganic bulk phase. Pores in nanoporous materials typically have a diameter of 100 nanometers or less. Thermal stability, chemical resistance, hardness, huge surface area, electron transport, conductance, and impedance are just a few of the physical and chemical features of nanoporous materials. The research community has been particularly interested in organic, inorganic, and hybrid porous materials over the last decade, owing to their wide range of promising applications in areas as diverse as (nano)filtration, separation techniques, heterogeneous supported catalysis, biomedical applications, template-assisted synthesis of nanomaterials, materials for phonic and thermic isolation, and patterned nanomaterials for microelectronics and photovoltaics.

This book presents the latest research on nanopores and nanoporous materials.

It provides the current state of nanopores technology as well as recent advances in their production, characterization, and application. The chapters present elegant approaches to functional nanoporous materials as well as their scope and limitations. Also discussed are important discoveries in the synthesis of nanoporous materials, ranging from soft porous materials to hard porous materials like porous metals and metal oxides, as well as significant advances in their applications to date. This book is a wonderful working resource for industrial scientists and engineers who want to take their knowledge of this unusual material to the next level and use it to drive technological innovation. Furthermore, it is a useful reference for academic researchers and graduate students working in the domains of biomass conversion, catalysis, materials science, green and sustainable chemistry, and chemical/process engineering. The goal of this book is to highlight the importance, functioning, and usability of nanoporous materials, as well as to encourage young and inquisitive minds to get involved in this field. We made every effort to incorporate as many relevant details as possible in this book, and we would like to express our gratitude to all of the authors who contributed significantly to it through their knowledge, efforts, and time. We also want to express our gratitude to IntechOpen for their assistance in publishing and bringing this book to life.

Sadia Ameen
Advanced Materials and Devices Laboratory,
Department of Bio-Convergence Science,
Advanced Science Campus,
Jeonbuk National University,
Republic of Korea

M. Shaheer Akhtar
New and Renewable Energy Material Development Center (NewREC),
Jeonbuk National University,
Republic of Korea

Hyung-Shik Shin
Korea Basic Science Institute (KBSI),
Gwahak-ro, Yuseong-gu, Daejon, Republic of Korea

Chapter 1

Introductory Chapter: Progress in Nanoporous Materials - An Introduction

Sadia Ameen

1. Introduction

The discovery of altering and prescribing properties of materials by controlling the size in the range of 1–100 nm has sparked interest in manufacturing materials from nanoscale building blocks. In accordance with the International Union of Pure and Applied Chemistry (IUPAC), the porous materials in macroscale range pose the pore sizes greater than 50 nm whereas, the mesoporous materials lie in the pore size range in between 2 and 50 nm, and the pore sizes less than ~2 nm belong to microporous materials. A unique subset of porous materials with the pore sizes ranges from 0.2 to 0.95 nm which basically called nanomaterials is popular field of considerations in many applications because they are presenting the special volume ratio of pore space to materials volume. Nanoporous materials are a type of nanostructured material that has a large specific surface area, a big pore volume, a uniform pore size, a rich surface chemistry, a significant porosity, and an ordered uniform pore structure. The well-ordered inorganic or organic frameworks support the regular, porous structure in nanoporous materials. Natural nanoporous materials are abundant in nature, but manmade nanoporous materials can also be created. These porous materials have the ability of allowing just certain compounds to pass through while blocking others [1]. The covalent organic frameworks, silicates, activated carbon, ceramics, zeolites, pillared materials, metal–organic frameworks, non-siliceous materials, aerogels, different polymers, and hybrid inorganic porous materials are examples of natural and manmade nanoporous solids. Long-range structural order or disorder can be found in nanoporous materials, which have pores ranging in size from a few nanometers to tens of nanometers. Nanomaterials including their excellent surface behavior such as high surface area and pore confinement effects are used in some applications, such as catalysis. New bottom-up techniques, such as molecular templating and intercalation, are required for the manufacturing and processing of porous materials in nanoscale range with programmable shapes and characteristics. Nanoporous materials have significant potential for the generation of new functional materials with better and tunable behaviors for use in various applications such as adsorption membranes, energy storage devices, sensors, different catalytic applications including photocatalysis, and biotechnology. The following are some of the applications of nanoporous materials that are studied in greater depth:

IntechOpen

2. Nanoporous materials for the manufacturing of clean energy and storage

Hydrogen as a clean energy carrier is essential for future energy supply. Fossil fuels, water electrolysis, and biomass can all be used to generate hydrogen. However, hydrogen must be created in a safe and environmentally friendly manner. Nanoporous material catalysts are essential for the effective conversion of coal to hydrogen at very low-cost and the capturing of carbon compounds. The development of fuel cells, in which hydrogen is the primary fuel and converts to power with water as a byproduct, is also critical. Nanoporous materials, such as carbon nanotubes, are critical in this process and offer considerable potential as future catalysts in fuel cells. Ameen et al. used a hydrothermal approach to synthesize well-crystalline porous cobalt oxide (Co_3O_4) nanorods (NRs) for use as an electrode material in supercapacitor applications. The porous and smooth morphology of Co_3O_4 NRs was synthesized at a low calcination temperature of 300°C, but the rod morphology changed to stacks of nanoparticles at a high calcination temperature of 500°C. The surface area and pore volume of porous Co_3O_4 NRs decreased as the calcination temperature increased, according to Emmett–Teller (BET) surface area study. The porous Co_3O_4 NRs calcined at 300°C as electrode was to manufacture pseudo-supercapacitors and achieved a specific capacitance of 226.3 Fg^{-1} (at scan rate = 10 mVs^{-1}). The capacitance value is superior to Co_3O_4 NRs-500°C electrode [2]. Porous cobalt oxide (Co_3O_4) nanocubes (NCs) have also been reported by Ameen and colleagues for use in electrochemical supercapacitors. Using cyclic voltammetry in KOH electrolyte, the capacitive characteristics of porous Co_3O_4 NCs electrodes were examined, and a high specific capacitance of ~430.6 F/g was reported at a scan rate of 10 mVs^{-1}. The porous Co_3O_4 NCs demonstrated outstanding structural stability during cycling, as well as promising capacity retention, implying that porous Co_3O_4 NCs are of superior quality as electrochemical supercapacitor electrodes [3].

3. Catalyst application of nanoporous materials

Chemical and fuel production have benefited greatly from heterogeneous catalysis. To boost catalytic activity and selectivity, more efficient catalytic procedures are required. As a result, customized catalytic materials with specified microstructures are essential. Nanoporous materials are very fastidious electro catalysts for the oxidizing small-scale organic molecules such as formic acid, acetic acid, methanol and ethanol due to their relatively large surface areas, prominent surface chemistry and small specific densities. Platinum is thought to be the finest catalyst for these reactions among pure metals. Although just a quarter of the pore surface area is electrochemically accessible, the mass specific activity is equivalent to platinum catalysts. The roughness factor, which govern the area of electrochemical surface to geometric portion, is a key element in determining catalytic performance. In a nanoporous environment, reactant molecules have a longer residence time, which permits them to linger longer inside the pores and have more opportunities to collide with the electrode surface than in a nonporous environment [4]. Ameen and colleagues synthesized ZnO-flowers photocatalyst to check the application in the crystal violet (Cv) dye degradation. Cv-dye degradation was particularly fast in the as-synthesized ZnO-flowers, with a degradation rate of 96% in 80-minute time interval [5]. Polyaniline/graphene (PANI–Gr) nanocomposites were made by in-situ polymerizing aniline monomer along with Gr in another study. The photocatalytic degradation of Rose Bengal (RB) dye was achieved using the nanocomposites as an effective photocatalyst. Gr was found in PANI–Gr nanocomposites with considerable interaction/bonding between PANI and Gr, as evidenced by the absorption characteristics. The imine (-NH) of PANI and the carboxylic group on the surface of Gr sheets formed a partial hydrogen bond in the PANI–Gr nanocomposites. Under light exposure, the produced PANI–Gr nanocomposites significantly degraded the RB dye by 56% within 3 h. The inclusion of Gr sheets in PANI–Gr nanocomposites might result in substantial charge separation of photogenerated electron–hole pairs under light irradiation, was related to the significant degradation of RB dye when compared to PANI [6]. The structural and surface characterizations of nanocomposites of poly(1-naphthylamine)/SiO_2 and poly(1-naphthylamine)/TiO_2 revealed an effective connection via hydrogen bonding between –NH group in PNA and –OH in nanomaterials (SiO_2/TiO_2). Under visible light illumination, the produced nanocomposites demonstrated considerable photocatalytic activity for the breakdown of methylene blue (MB) dye. Due to the presence of efficient charge separation of photogenerated e^-–h + pairs, PNA/TiO_2 nanocomposites posed the superior MB dye degradation of 60% as compared to PNA/SiO_2 (28%) and pure PNA (9%) [7].

4. Sensor materials made of nanoporous materials

Nanoparticles and nanoporous materials have a vast surface area and are extremely sensitive to environmental changes. Sensor made of these materials are frequently employed. The sensitivity of gas sensors is determined by their surface areas, and gas sensors made of nanoporous metal oxides like TiO_2 or ZnO are being manufactured and employed to flammable gas detectors. In general, the gas sensors detect changes in electric resistivity as a function of gas concentration, and their sensitivity is proportional to surface area. Ameen et al. described a low-temperature solution method for fabricating aligned nanoporous ZnO NRs on FTO glass and using them as an electron mediator to fabricate a highly sensitivity chemical sensor for p-nitrophenylamine (p-NPA) detection in an aqueous buffer electrolyte. A high and repeatable sensitivity of ~184.26 μA mM^{-1} cm^{-2} and a quick response time of 8 s is attained by p-NPA chemical sensor based on aligned nanoporous ZnO NRs electrode. With a correlation efficiency of R = ~0.97569, the manufactured p-NPA chemical sensor had a respectable detection limit of ~53.7 M and good linearity in the region of 5–20 M [8]. Ameen and colleagues reported a modified electrode of poly (1-naphthylamine) nanoglobules for the detection of different alcohols using fabricated ultra-high sensitive chemical sensors. With a reaction time of 10 seconds, the constructed ethanol chemical sensor based on PNA nanoglobules has a high and repeatable sensitivity of ~1.66 $\mu A mM^{-1}$ cm^{-2}. The developed PNA nanoglobules-based chemical sensor had a linear dynamic range (LDR) of 0.78 mM to 50 mM, with a correlation efficiency of R = ~0.965 [9].

5. Nanoporous materials in biological applications

Because nanoporous materials are biocompatible, they can be used to create enzymatic nanomaterials which normally mimic various biological reactions. Enzymes immobilized on nanoporous materials can be utilized in biological reactors to generate pharmaceuticals, decontaminate waste, and other applications. Biosensors can be made from nanoporous materials. The electrochemically produced nanocages-augmented PANI nanowires (NCa-PANI NWs) on silicon (Si) substrate were used to create a non-enzymatic glucose biosensor. Current (I) – voltage (V), cyclic voltammetry, and amperometry measurements were used to evaluate the sensing parameters for the NCa-PANI NWs electrode. The sensing findings demonstrated that the manufactured non-enzymatic sensor responded well to glucose, with a stable, dependable, and high sensitivity value = ~156.4 mA mM^{-1} cm^{-2}, a promising limit of detection = ~0.657 M, and R = ~0.99493 [10].

6. Drug delivery using nanoporous materials

Because of its great porosity and surface area, porous silicon is an ideal material for drug delivery. Small compounds like doxorubicin have been placed into porous silicon pores as therapeutic agents. For drug delivery applications, various factors must be taken into account, including pore size, as the substance must smaller to the pore diameter for traversing the pores. Porosity is significant because the amount of medicine loaded into the pores is proportional to the capacity of the pores. This book is a valuable source of nanoporous materials for researchers interested in inventing and employing procedures to synthesis, characterize, and model nanopores in the general areas of materials science, chemical engineers, biotechnology, nanobiotechnology, biomedical engineers, and electrochemistry. This book presents the most up-to-date research on nanopores and nanoporous materials, with a particular emphasis on practical analytical applications of nanoporous materials. The major goal of this book is to provide readers in academia, industry, engineering, and biomedical disciplines with a comprehensive professional reference on nanopores. Moreover, this book provides comprehensive knowledge to readers about nanoporous materials and presents the most recent advances in a variety of domains, including synthesis, characterization, and surface modification, as well as adsorption and separation processes, biological and catalytic applications. Fundamentally, this book comprises chapter on key topics such as nanoporous materials synthesis, characterization methodologies, nanomaterials surface modification or surface functionalization, designing of novel catalyst, and nanostructure tailoring.

Author details

Sadia Ameen
Advanced Materials and Devices Laboratory, Department of Bio-Convergence Science, Jeonbuk National University, Republic of Korea

*Address all correspondence to: sadiaameen@jbnu.ac.kr

References

[1] Olister, Paul. "Nanoporous Materials" Cientifica. Retrieved 6 February 2013.

[2] G.S. Jang, S. Ameen, M. S. Akhtar, E. Kim, H. S. Shin, Chemistry Select 2 (2017) 8941.

[3] G.S. Jang, S. Ameen, M. S. Akhtar, H. S. Shin, Ceramics Intern. 44 (2018) 588.

[4] Losic, Dusan, Santos, Abel (2010) Electrochemically Engineering Nanoporous Materials, Springer Series in Material Science 220: 1-337.

[5] S. Ameen, M. S. Akhtar, M. Nazim, H. S. Shin, Materials Letters 96 (2013) 228.

[6] S. Ameen, H. K. Seo, M. S. Akhtar, H. S. Shin, Chem. Eng. J. 210 (2012) 220.

[7] S. Ameen, M. S. Akhtar, Y. S. Kim, H. S. Shin, Appl. Catal. B: Environ. 103 (2011) 136.

[8] S. Ameen, M. S. Akhtar, H. K. Seo, H. S. Shin, Appl. Catal. s A: Gen. 470 (2014) 271.

[9] S. Ameen, M. S. Akhtar, A. Umar, H. S. Shin, Chem. Eng. J. 229 (2013) 267.

[10] S. Ameen, M. S. Akhtar, H. S. Shin, Appl. Catal. A: General 517 (2016) 21.

Chapter 2

Potential Application of Nanoporous Materials in Biomedical Field

Saraswati Prasad Mishra, Shweta Dutta, Anil Kumar Sahu, Koushlesh Mishra and Pankaj Kashyap

Abstract

Nanoporous materials are the substances having pores of size 100 nanometers in a frame work organic or inorganic substance. These substances are used in medical devices such as bioartificial organ and biosensing. Nanoporous material has also importance in the field of diagnostics. This chapter basically explains about the nanoporous material in detail along with its types. The methods of fabrication of these nanoporous material area also explained. The chapter also deals with the characterization of the materials. Moreover present application of nanoporous material such as in the field of biomedicals along with the future prospects is explained in the present chapter.

Keywords: nanoporous material, biosensing, organic, inorganic, medical device

1. Introduction

Nanoporous material is a structure containing framework of organic or inorganic substances having pores of size 100 nanometers. The pores found in nanoporous material contain either gas or liquid filled in it. A Nanoporous material is used recently in novel medical devices, implants or making bioartificial organs and biosensing. Advancement in the field of nanofabrication made it possible to produce nanoporous material with desired size of pores, distribution of pores in the nanoporous material as well as their porosity and chemical nature. Eventually it made the nanoporous material more attractive to carry out process of regulation and transportation at the molecular level. Basically nanoporous material is used for size sorting; antibiofouling behavior along with it is used in medical devices as mentioned above. In near future it is possible that nanoporous material can be functionalized with smart polymers that can initiate or modulate transportation at bio-molecular level in response to different kind of stimuli such as ion, change in pH or temperature [1]. This can eventually help in development of such medical device that can act in accordance to the changing physiological needs. The body cells naturally have proteins of nano-size that helps in regulating movement of biomolecules across the membranes. In similar way nanoporous material functionalized with smart polymer will differentiate between the biomolecules that has to be transported from the biomolecules that are not to be transported. As nanoporous material have small pore size but contains a larger surface porosity it becomes ideal to be used in activities like ion exchange, catalysis, sensing [2–4]. Nanoporous material has an important role to play diagnostic field as it is used in combinatorial biochemistry on-a-chip, in analysis of DNA, in activity like cell manipulation and chromatography as well [5, 6]. Moreover it can also be used in boosting devices that are used to store energy as nanoporous material shows a greater conductivity to electrolytes. The present chapter explains about the nanoporous materials, their relevance in present day as well as their future prospects and their classification. The chapter also elaborates about the fabrication methods, nanopores techniques along with the characterization of nanoporous material and their applications [7–10].

2. Types of nanoporous materials

Nanoporous material are generally grouped into two class i.e. bulk material and membranes. Under bulk material activated carbon and zeolites are the examples whereas when membranes are concerned then cell membrane is an example of nanoporous membrane. Nanoporous materials are made using a chemical reagent that is basically inorganic and a structure is provided by using the organic templates. It can be said that nanoporous material is made by polymerization of inorganic monomers that are assiated by the templates of organic molecules. Many nanoporous material are also made by using minerals instead of chemicals reagent as inorganic source. In case of mineral nanoporous material templating is based on the initial structure of the mineral itself [11, 12].

2.1 Classifications

Nanoporous materials can be of different types as discussed above. Below are classification of nanoporous material based on pore size and the network material used.

2.1.1 Classication by pore size

The pores of nanoporous material vary from 1 nm to 1000 nm. In accordance to IUPAC there are following class of nanoporous material [11–13].

- **Microporous material:** The pore size in this kind of material is between 0 to 2 nm

- **Mesoporous material:** In mesoporous material the pore size ranges from 2 to 50 nm

- **Macropores material:** Macroporous materials are those materials where the pores size is greater than 50 nm.

Comparison between these three pore systems is given in **Figure 1**. There is no order found between above mentioned pore materials, mostly they are random in nature (**Table 1**) [14–17].

2.1.2 Classication based on network material

In the field of nanoporous material one of the most important thing is to have network material of desired chemical composition. These network materials can be classed into two categories

i. Organic material

ii. Inorganic material

One of the most important goals in the field of nanoporous materials is to achieve any possible chemical composition in the network materials "hosting" the pores. It makes sense to divide the materials into two categories:

i. Organic materials

ii. Inorganic materials.

i. Organic material:

As organic material act as template for the inorganic material to form the structure so it is the smaller group used nanoporous material. Different kinds of polymers are used under this category [18, 19].

ii. Inorganic Material: Inorganic material are the main group in nanoporous material. Following nanoporous materials are used [20].

- Inorganic oxide type materials such as porous silica or porous titania or porpous material of zirconia is used.

- Nanoporous carbon materials are also used where active carbons are used. Mesoporous carbon materials are example under these groups.

- Sulphide and nitrites are also used under inorganic material. An AlPO4 material also comes under this.

3. Fabrication methods

The area of fabrication in materials of nanostructure is ever improving area with involvement of innovative techniques that helpful to different field of research and

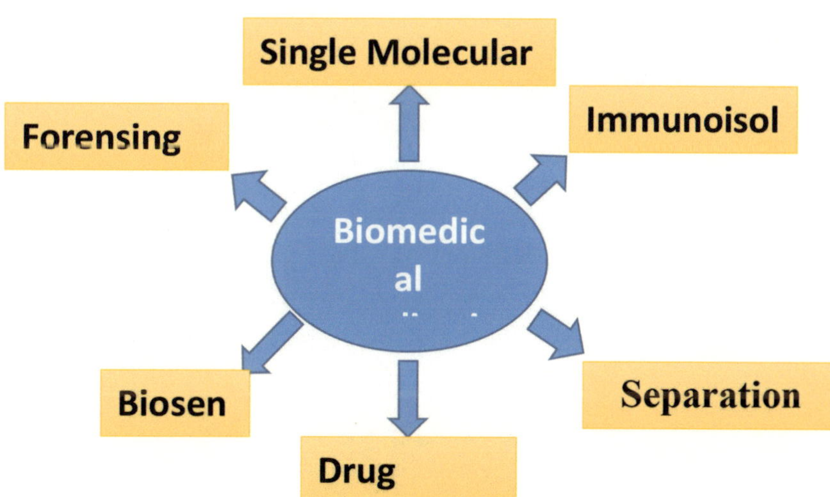

Figure 1.
Biomedical applications of nanoporous materials.

CONTENT	POLYMERIC	CARBON	GLASS	ALUMINOSILICATE	OXIDASE	METAL
PORE SIZE	Meso-macro	Micro-meso	Meso-macro	Micro-meso	Micro-meso	Meso-macro
SURFACE AREA / POROSITY	Low 0.6	High-0.3–0.6	Low-0.3–0.6	High0.3–0.6	Medium 0.3–0.6	Low 0.1–0.7
PERMEABILITY	Low- medium	Low- medium	High	Low	Low- medium	High
STRENGTH	Medium	Low	Strong	Weak	Weak- medium	Strong
THERMAL STABILITY	Low	High	Good	Medium- high	Medium- high	High
CHEMICAL STABILITY	Low –medium	High	High	High	Very high	High
COST	Low	High	High	Low medium	Medium	Medium
LIFE	Short	Long	Long	Medium long	Long	Long

Table 1.
That table contains various properties of nanoporous.

development [21]. Improvement seen in the field of nanofabrication and the growing interest in the domain of nano-manufacturing can help in the enhancement in methods of ultrafiltration [22]. Ideal properties of a protein sieve or a molecular membrane is that it contains uniformly distributed pores on an ultrathin membrane and that is fabricated in such a way that can be used in scalable and robust manner. It should be cleanable and reusable after sterilization. In this fabrication method, nanoporous membrane are made with ratio of pore size to thickness is around one. The said ratio between pore size to thickness helps in effective mass transportation due to enhanced selectivity and permeability. Fabrication of membrane is done at very lost cost so that it is scalable enough to have manufacturing at large scale. The defects that are seen during fabrication of membrane are pore size variation and absence of pores in membrane. As far as ultrafiltration is concerned absence of pore size is not that important and optimization of variation in pore size can be performed to have better functioning of membrane and optimum efficiency.

3.1 Nanopore techniques

Nanopores are nothing but pores having size in nanometer. They can be made either by using proteins that can form pores or by creating pores of nanosize in molecules. When nanopores are coated with iron and are present in a membrane which is electrically insulating act as single molecule identifier. Additionally it also acts as network of biological protein in bilayer of phospholipid. Nanopore technology is used as a detector for detecting the biological and chemical agent in nanoscale at molecular level. By the use of principle of electrophoresis a device based on nanopores pulls the molecules through nanopores into the solution and detect the molecule and ascertain their competence at analytically. Characterization of nucleic acid polymer is done in narrow and confined space in the nanopores. Nanopore sequencing technique has made DNA sequencing inexpensive and fast as characterization of single stranded DNA and RNA without labelling and amplification of it [23]. As nanopores are highly sensitive that lead to many research that helps in analyzing nucleic acid [24, 25].

3.2 Biological nanopores

Proteins are also capable of forming nanopores [26]. This kind of protein are typically have a structure like mushroom and the core of the mushroom shaped structure has hollow in it. Examples of some proteins capable of pores are α hemolysin, Phi 29 connector and MspA porin. The most initial biological nanopore is α hemolysin (α -HL) which is used in the area DNA sequencing. α-HL is produced from bacterium *Staphylococcus aureus* as an exotoxin. The specification of much room shaped protein is 232.4 kDa of transmembrane channel with a cap of diameter around 3.6 nm and barrel of 2.6 diameter barrel [27]. Then it is inserted in lipid bilayer and then manipulation is done [28].

3.3 Solid-state nanopores

These kinds of nanopores are made from silicon film, mostly silicon nitride. Various techniques are employed for solid state nanopores manufacturing which involves “fabrication by electron beam” and “Deploying and sculpting with ion beam” [29]. Solid nanopores have diameter ranging from sub nanometers to nanometers in hundreds and the change in diameter is based on the requirement of experimental parameter. SiN used in manufacturing of solid state nanopores shows better chemical and thermal stability as compared to lipid membrane [30].

Nanopores made of graphene expressed chemical properties that are unique and shows btter gains over the biological complements [31]. Solid state nanopores created many paths for research especially in DNA sequencing. Identifications of protein interaction nanofluidic device assembly. Solid state nanopores are suitable substitute for biological nanopores due to the unique chemical properties. Various measurement technique such as electronic and optical measurement are compatible with solid state nanopores. Reecent nanopores fabrication techniques are membrane technology for ion tracking [32, 33]. Production of metallic surfaced oxidative film ionic beam sculpting.

3.4 Anodic oxidation method on the metal aluminum

When electrochemistry and electrophysiology of anodic oxidation of metals was observed it resulted in fabrication of nanoporous oxides of metals that are self-ordering. Metals included are anodized form of aluminum oxide, nanotubular titania oxide and silicon [34]. The reasons due to which the anodic alumina oxide stands out are its hardness, high surface area and stability it shows chemically and thermally [35]. Selective metals such as Al, Nb, Ti, are studied for ordering behavior during the process of anodic oxidation. These metals are known as valve element [36]. Factors responsible for enhancement of the process are electrolyte type, its pH as well as concentration, temperature, surface and the voltage and current applied [37, 38].

3.5 Ion track-etching technology

This technology is used for generation of pore in materials that are insulating in nature. Several polymers are used to produce filtration films. The underlying principle is that when a material comes in the path of straight ion, due to penetration by high energy heavy ion a pore is seen in the material. By the help of appropriate reagent etching is done to enlarge the pores. Pore size can be made of dimension of nanometers to micrometers and cylindrical pores as well [39, 40].

To have a uniform etching surfactant are added during the process of ion track etching [41].

While using surfactants following few things are to be taken into consideration.

- When surfactant used gets adsorbed on the surface it tends to change susceptibility to chemical attack.
- Size of surfactant molecule is quite small in nanometer range [42].

3.6 Ion-beam sculpting

Ion beam sculpting has been matter of interest for the researchers for the meeting the challenges of nanopores. As it has low rate of shattering of ions, it gives better firmness and patterning of substrate which makes it crucial in meeting nanopore challenges High resolution of focused ion beam offers nanometer based sculpting [43].

3.7 Ion current rectification

Specific kind of transportation effect has been seen in nanocapillaries or nanopores having uneven shape and the reason being the nanosize of the opening. It is seen that there is rectification of ion current in this kind of nanopores whereas pH of electrolyte and the concentration remains the same. For the purpose of observation of rectification current voltage curves are used [44, 45]. Ion current rectification is behavior seen in many nanoporous system. A biological nanopore as well as artificial nanopores shows rectifying behavior [46, 47].

3.8 Electron-beam fabrication

Fabrication of solid state nanopore with small diameter is difficult. It is almost impossible to fabricate the nanopores which are less than 30 nm in terms of shape and size. By use of FIB nanopres can be etched but due to low etch rate limitation on film thicknesss can be seen [48–50]. Nanopores can be significantly condensed to almost 10 nanometer from 50 to 100 nanometrs by use of ion beam or electron beam having high energy. Solid state nanopores are very effective in detection of single molecule when pore diameter is as equal as molecule diameter [51, 52].

4. Characterization

4.1 FTIR spectroscopy

Fourier transform infrared spectroscopy (FTIR) is a type of spectroscopy that concerned with the infrared portion of the electromagnetic spectrum that helps in identifying a compound by investigating the composition of a sample. Specific frequencies of Infra-red (IR) radiation is absorbed by molecule based the functional group present in it [53].

4.2 Raman spectroscopy

It is a type of vibrational spectroscopy at molecular level which originated as inelastic light scattering process. In this spectroscopy sample molecules scatters a laser photon and there will be gain or loss of energy. Energy lost is indicator of change in energy or wavelength of the laser photon. Energy lost is characteristic to a specific bond in molecule. With Raman spectroscopy an exact spectral fingerprint can be obtained specific to molecule or any molecular structure [54].

4.3 UV-Vis spectroscopy

UV–Vis spectroscopy is different from earlier two as it is concerned with electronic transition occurring within a molecule. When a continuous striking of radiation is done on a molecule then some portion of the radiation get absorbed and the remaining radiation is passed across a prism it gives spectrum that has gap in between. This spectrum is called as absorption spectrum and due to absorption of energy there is transition of molecule from low energy to higher energy state [55].

4.4 Energy-dispersive X-ray spectroscopy (EDX)

These spectroscopies are used for analysis of element and determine the characteristics of chemical aspect of sample. X-ray is a form of energy released when sample is being bombared with high energy beam that leads to ejection of excited electron from inner shelf creating a hole and the hole formed is filled by electron from a high energy outer cell and during this energy. To measure the X-ray in terms of number and energy the instrument used is energy-dispersive spectrometer. X-ray helps in determining composition of element in a specimen [56].

4.5 X-ray diffraction (XRD)

X-ray Diffraction (XRD) is a technique which studies the diffraction produced by X-ray through the lattice and determines the characteristics of lattice. It helps in determine structure of zeolite. The sample preparation for this technique is easy and the pace of the test is quick [57].

4.6 Scanning electron microscope (SEM)

Scanning electron microscope (SEM) is an instrument that is different from normal microscope as it makes image by using electrons rather than light. In SEM when scanning of sample is done by the beam of primary electron, the surface electrons get excited and that leads to release or emission secondary electron from the surface that results in formation of image. SEM is capable of producing images having high resolution that enablkes the observer to examine the close features with higher magnification. The images formed from SEM gives details about particle size and surface of sample [58].

4.7 Transmission electron microscope

In TEM utilizes the electron beam that has transmitted partially across a very thin specimen. This beam helps in getting the image. TEM helps in determining or acquiring information about structure and particle size of the sample under study. TEM is slightly better in magnitude than SEM [59].

4.8 Nitrogen adsorption/desorption isotherms

This technique is used for determination of characteristics of surface zeolite. It provides information related to the entire surface such as internal, external along with the diameters of mesopores [60].

5. Biomedical applications of nanoporous materials

Applications of nanoporous materials in biomedical field has been explored and discovered and there are many more under exploration still to be discovered. Nnaoporous membranes act as semipermeable membrane or compartment in many implantable devices that keep the drug or the implant and allow the passage of desired molecule. Moreover nanoporous material has application in variety of biomolecular application. It is also used in field of diagnosis and separation of protein [61].

5.1 Separation and sorting of biomolecules

Sorting or separation is essential to purify and isolate the molecules from the stream of biological feed. This application has a huge importance in the industry like pharmaceutical manufacturing, biotechnology and food industry. Currently techniques like gel electrophoresis or size exclusion chromatography are relevant and used in separation science [62, 63]. Examination of biomolecular separation in pores which are more ordered has been done recently. Synthetic nanoporous membrane has been used as support system for the cells of kidney as they filter blood and retain proteins present in serum and filter out the waste materials [64]. The material that flow through the nanoporous material can be regulated externally [65].

5.2 Biosensing

Proteins pores that are membrane bound are used by sensory system as a detector of stimuli and facilitate the cells to respond accordingly. Biosensing has its application in fields like pharmaceutical industry, in the sector of medical diagnosis and it is also used for detecting of hazardous biomolecules. In these applications there is combination of physiochemical detection component with biological component for detection of analytes in stream of biological feed. Sensory systems use a variety of membrane-bound protein pores to detect molecules and facilitate cells to respond to stimuli. Such biosensing is also important in many technological areas including pharmaceutical industry, medical diagnosis, and detection of hazardous biomolecules. In a majority of these applications the biosensing device combines a biological component with a physiochemical detection component to detect analytes in biological feed streams [66].

5.3 Single molecular analysis

Nanoporous materials are also used probing of biomacromolecules such as DNA, RNA, and proteins one by one for single-molecule analysis. Information of biomacromolecules such as concentration, sequence, size or structure can be accessed by measurement of magnitude, frequency and blockage duration of ion current when the biomolecules are passed through the nanopore which is embedded in insulating membrane [67]. Earlier research in the field of single molecule analysis had utilized lipid membrane that had been incorporated in polymeric films like Teflon having aperture of microsize. The only drawback with micro-sized pores having polymeric support is rupture of lipid membrane after a small period of use and this technique has to be improved to have better durability. But nanoporous membrane shows better result in supporting protein pores in the process of single molecule analysis [68].

5.4 Immunoisolation

Immunoisolation means to protect implanted cells or the drug release systems from any kind of an immune reaction. It is done by encapsulating the implanted cell or drug in a nanoporous semipermeable membrane. This nanoporous material isolate the encapsulated drug or cell from the immune system of body. The pores allows entry of glucose, insulin and oxygen to pass through but it is impearmable to immunoglobulins. Only requirement for nanoporous material to use in immunoisolation is that it should be compatible foul resistant for *in vivo* functions [69].

5.5 Drug delivery

In vivo delivery systems are developed to supply of drugs in a controlled manner where it is needed. Controlled delivery system is used to deliver drugs in effective way so as to eliminate any kind of improper dosing. Nanoporous membranes having controlled pore size, desired membrane thickness and porosity can deliver controlled release drugs in capsule form [70]. By coupling it with biosensors a smart drug delivery systems can be developed that will respond according to the physiological conditions.

5.6 Forensing analysis

Nanoporous gold (NPG) being a good conductor and having suitable pore-size distribution with large surface area, and can enhance the electrochemical response to the enzymatic substrates namely NADH and H2O2 depending on their low coordinated Au atoms. All said advantages make it perfect for construction of dehydrogenase- and oxidase-based biosensors which will show improved sensitivity and anti-interference ability. DNA sensor which is based on an NPG electrode and is prepared by the process of dealloying Ag from Au/Ag alloy and multifunctional encoded AuNP. The active surface area of the NPG electrode is 9.2 times larger as compared to bare flat as characterized by CVs. Fabrication of DNA biosensor was done by immobilizing capture-probe DNA on the NPG electrode and hybridization with target DNA, which further hybridized with the reporter DNA loaded on the AuNP. The AuNP contained two kinds of bio bar-code DNA, one complementary to the target DNA, while the other was not, reducing the cross reaction between the targets and reporter DNA on the same AuNP. Besides DNA detection, NPG is also used in making an amperometric immunosensor [71–73].

6. Application

Nanoporous materials can enhance the performance devices used in biomedical field such as immunoisolation devices, devices used for dialysis, targeted drug delivery systems, bioanalytical devices, and biosensors. The main properties that nanoporous membranes should have so that it can be used in biomedical applications are having a pore size of a few tens of nanometers or below it and the pore size distribution should be in order that help us to achieve high biomolecule selectivity; high porosity as well as low thickness in order to enable high analyte flux; mechanical stability; and chemical stability [74]. The central issue of membrane is Pore geometry, biofouling resistance, and biocompatibility so that it can be used like interfaces in implantable devices. Porous material has become a potential drug delivery system for lots of biomedical application. They can be modified internally as well as externally to load the required molecule efficiently. Moreover outer layer can acts as a barrier and help in delaying the release of drug. Porous material has many advantages over the prominently used organic material for the drug delivery. They show better stability, better loading capacity, and provide better protection to the loaded material from degradation. Although porous material has potential to used but the obstacle is how it can be transferred to the clinic successfully [75].

7. Future prospects

Porous materials are the materials of future as they show many advantages over the prominent materials used in recent times. They provide versatile porosity and the pore size can be tailored according to the need. It also has better drug loading capacity. With all the said advanatges nanopores material can be in demand in future in many fields.

Author details

Saraswati Prasad Mishra[1*], Shweta Dutta[2], Anil Kumar Sahu[2], Koushlesh Mishra[3] and Pankaj Kashyap[2]

1 RITEE College of Pharmacy, Raipur, Chhattisgarh, India

2 Royal College of Pharmacy, Raipur, Chhattisgarh, India

3 Raigarh College of Pharmacy, Raigarh, Chhattisgarh, India

*Address all correspondence to: saraswatim3@gmail.com

References

[1] Adiga SP, Curtiss LA, Elam JW, Pellin MJ, Shih CC, Shih CM, Lin SJ, Su YY, Gittard SD, Zhang J, Narayan RJ. Nanoporous materials for biomedical devices. Jom. 2008 Mar 1;60(3):26-32.

[2] J. Ly, M. Alexander, and S.E. Quaggin, Current Opin- ion in Nephrology and Hypertension, 13 (2004), pp. 299-305.

[3] D.A. LaVan, T. McGuire, and R. Langer, Nature Bio- technology, 21 (2003), pp. 1184-1191.

[4] *L. Leoni*, A. Boiarski, and T.A. Desai, Biomedical Mi- crodevices, 4 (2002), pp. 131-139.

[5] Z. Huang et al., Journal of Medical Devices, 1 (2007), pp. 79-83.

[6] F. Martin et al., Journal of Controlled Release, 102 (2005), pp. 123-133.

[7] M. E. Davis, "Ordered porous materials for emerging applications," Nature 417, 813-821 (2002).

[8] I. Moriguchi, M. Honda, T. Ohkubo, Y. Mawatari, and Y. Teraoka, "Adsorption and photocatalytic decomposition of methylene blue on mesoporous metallosilicates," Catal. Today 90, 297-303 (2004).

[9] H. Yamada, H. Nakamura, F. Nakahara, I. Moriguchi, and T. Kudo, "Electrochemical study of high electrochemical double layer capacitance of ordered porous carbons with both meso/macropores and micropores," J. Phys. Chem. C 111, 227-233 (2007).

[10] L. L. Zhang and X. S. Zhao, "Carbon-based materials as supercapacitor electrodes," Chem. Soc. Rev. 38, 2520-2531 (2009).

[11] Polarz S, Smarsly B. Nanoporous materials. Journal of nanoscience and nanotechnology. 2002 Dec 1;2(6):581-612.

[12] Ng EP, Mintova S. Nanoporous materials with enhanced hydrophilicity and high water sorption capacity. Microporous and Mesoporous Materials. 2008 Sep 1;114(1-3):1-26.

[13] J. Rouqerol, D. Avnir, C. W. Fairbridge, D. H. Everett, J. H. Haynes, N. Pernicone, J. D. Ramsay, K. S. W. Sing, and K. K. Unger, *Pure Appl. Chem.* 66, 1739 (1994).

[14] K. S. W. Sing, D. H. Everett, R. A. W. Haul, L. Moscou, R. A. Pierotti, J. Rouquérol, and T. Siemieniewska, *Pure Appl. Chem.* 57, 603 (1985).11 12:

[15] D. W. Schaefer and K. D. Keefer, *Phys. Rev. Lett.* 53, 1383 (1984).

[16] Á. Kukovecz, Z. Kónya, I. Pálinkó, D. Mönter, W. Reschetilowski, and I. Kiricsi, *Chem. Mater.* 13, 345 (2001).

[17] K. D. Keefer and D. W. Schaefer, *Phys. Rev. Lett.* 56, 2376 (1985).

[18] D. W. Schaefer, *Science* 243, 1023 (1989).

[19] N. Koshida and B. Gelloz, *Curr. Opin. Colloid Interface Sci.* 4, 309 (1999).

[20] K. L. Kavanagh and M. J. Sailor, *Science* 255, 66 (1992).30 31

[21] H. P. Hentze and M. Antonietti, *Curr. Opin. Solid State Mater. Sci.* 5, 343 (2001).

[22] Mireles M, Gaborski TR. Fabrication techniques enabling ultrathin nanostructured membranes for separations. Electrophoresis. 2017 Oct;38(19):2374-88.

[23] Montagne F, Blondiaux N, Bojko A, Pugin R. Molecular transport through

nanoporous siliconnitride membranes produced from self-assembling block copolymers. Nanoscale. 2012;4(19):5880-6.

[24] Rhee M & Burns MA (2006). Nanopore sequencing technology: research trends and applications. *Trends Biotechnol* 24(12): 580-586.

[25] Healy K, Schiedt B & Morrison AP (2007). Solid-state nanopore technologies for nanopore-based DNA analysis. *Nanomedicine (Lond)* 2(6): 875-897.

[26] Tobkes N, Wallace BA & Bayley H (1985). Secondary structure and assembly mechanism of an oligomeric channel protein. *Biochemistry* 24(8): 1915-1920.

[27] Storm AJ, Chen JH, Ling XS, Zandbergen HW & Dekker C (2003). Fabrication of solid-state nanopores with single-nanometre precision. *Nat Mater* 2(8): 537-540

[28] Song LZ, Hobaugh MR, Shustak C, Cheley S, Bayley H, Gouaux JE (1996). Structure of staphylococcal alpha-hemolysin, a heptameric transmembrane pore. *Science* 274: 1859-1866.

[29] Briggs K, Madejski G, Magill M, Kastritis K, de Haan HW, McGrath JL & Tabard- Cossa V (2018). DNA Translocations through Nanopores under Nanoscale Preconfinement. *Nano Lett* 18(2): 660-668.

[30] Bayley H (2009). Membrane-protein structure: Piercing insights. *Nature* 459(7247): 651-652.

[31] Thompson JF & Milos PM (2011). The properties and applications of singlemolecule DNA sequencing. *Genome Biol* 12(2): 217.

[32] Meyers S, Downing JR & Hiebert SW (1993). Identification of AML-1 and the (8;21) translocation protein (AML-1/ETO) as sequence-specific DNA-binding proteins: the runt homology domain is required for DNA binding and protein-protein interactions. *Mol Cell Biol* 13(10): 6336-6345.

[33] Abgrall P & Nguyen NT (2008). Nanofluidic devices and their applications. *Anal Chem* 80(7): 2326-2341.

[34] Anglin EJ, Cheng L, Freeman WR & Sailor MJ (2008). Porous silicon in drug delivery devices and materials. *Adv Drug Deliv Rev* 60(11): 1266-1277.

[35] Ghicov A & Schmuki P (2009). Selfordering electrochemistry: a review on growth and functionality of TiO2 nanotubes and other self-aligned MO(x) structures. *Chem Commun (Camb)* (20): 2791-2808.

[36] Chu YY, Wang WJ & Wang M (2010). Anodic oxidation process for the degradation of 2, 4-dichlorophenol in aqueous solution and the enhancement of biodegradability. *J Hazard Mater* 180(1-3): 247-252.

[37] Ren Y, Ma Z & Bruce PG (2012). Ordered mesoporous metal oxides: synthesis and applications. *Chem Soc Rev* 41(14): 4909-4927.

[38] Yilmaz S, Uslu B & Ozkan SA (2001). Anodic oxidation of etodolac and its square wave and differential pulse voltammetric determination in pharmaceuticals and human serum. *Talanta* 54(2): 351-360.

[39] Kumeria T, Santos A & Losic D (2014). Nanoporous anodic alumina platforms: engineered surface chemistry and structure for optical sensing applications. *Sensors (Basel)* 14(7): 11878-11918.

[40] Li F, Guijt RM & Breadmore MC (2016). Nanoporous Membranes for

Microfluidic Concentration Prior to Electrophoretic Separation of Proteins in Urine. *Anal Chem* 88(16): 8257-8263.

[41] Fleischer RL, Price RB & Walker RM (1969). Nuclear tracks in solids. *Sci Am* 220(6): 30-39.

[42] Yuan Z, Wang C, Yi X, Ni Z, Chen Y & Li T (2018). Solid-State Nanopore. Nanoscale *Res Lett* 13(1): 56.

[43] Giselbrecht S, Gottwald E, TruckenmuellerR,TrautmannC,WelleA, Guber A, Weibezahn KF et al (2008). Microfabrication of chip-sized scaffolds for three-dimensional cell cultivation. *J Vis Exp* (15): e699.

[44] Ali M, Ramirez P, Duznovic I, Nasir S, Mafe S & Ensinger W (2017). Label-free histamine detection with nanofluidic diodes through metal ion displacement mechanism. *Colloids Surf B Biointerfaces* 150: 201-208

[45] Oda K, Csige I, Henke RP & Benton EV (1992). A new method for internal calibration of nuclear track detectors. *Int J Rad Appl Instrum D* 20(3): 505-510.

[46] Ali M, Ramirez P, Nasir S, Nguyen QH, Ensinger W & Mafe S (2014). Current rectification by nanoparticle blocking in single cylindrical nanopores. *Nanoscale* 6(18): 10740-10745.

[47] Tong X, Aoyama H, Tsukihara T & Bai D (2014). Charge at the 46th residue of connexin 50 is crucial for the gap-junctional unitary conductance and transjunctional voltage-dependent gating. *J Physiol* 592(23): 5187-5202.

[48] Dondapati SK, Kreir M, Quast RB, Wustenhagen DA, Bruggemann A, Fertig N & Kubick S (2014). Membrane assembly of the functional KcsA potassium channel in a vesicle-based eukaryotic cell-free translation system. *Biosens Bioelectron* 59: 174-183.

[49] Yamamoto T & Doi M (2014). Electrochemical mechanism of ion current rectification of polyelectrolyte gel diodes. *Nat Commun* 5: 4162.

[50] Walz MM, Schirmer M, Vollnhals F, Lukasczyk T, Steinruck HP, Marbach, H (2010). Electrons as "invisible ink": fabrication of nanostructures by local electron beam induced activation of SiOx. *Angew Chem Int Ed Engl* 49(27): 4669-4673.

[51] Zhu C, Du D, Eychmuller A & Lin Y (2015). Engineering Ordered and Nonordered Porous Noble Metal Nanostructures: Synthesis, Assembly, and Their Applications in Electrochemistry. *Chem Rev* 115(16): 8896-8943.

[52] Fleischer R.L, Turner LG, Paretzke HG & Schraube H (1984). Personnel neutron dosimetry using particle tracks in solids: a comparison. *Health Phys* 47(4): 525-531.

[53] Noh JH, Nikiforov M, Kalinin SV, Vertegel AA & Rack PD (2010). Nanofabrication of insulated scanning probes for electromechanical imaging in liquid solutions. *Nanotechnol* 21(36): 365302.

[54] Sumikama T (2016). Origin of the Shape of Current-Voltage Curve through Nanopores: A Molecular Dynamics Study. *Sci Rep* 6: 25750.

[55] Cai L, Song AY, Wu P, Hsu PC, Peng Y, Chen J, Liu C, Catrysse PB, Liu Y, Yang A, Zhou C. Warming up human body by nanoporous metallized polyethylene textile. Nature communications. 2017 Sep 19;8(1): 1-8.

[56] Qi J, Motwani P, Gheewala M, Brennan C, Wolfe JC, Shih WC. Surface-enhanced Raman spectroscopy with monolithic nanoporous gold disk substrates. Nanoscale. 2013;5(10):4105-9

[57] Cai J, Kimura S, Wada M, Kuga S. Nanoporous cellulose as metal nanoparticles support. Biomacromolecules. 2009 Jan 12;10(1):87-94.

[58] Indira K, Mudali UK, Rajendran N. Corrosion behavior of electrochemically assembled nanoporous titania for biomedical applications. Ceramics International. 2013 Mar 1;39(2):959-67.

[59] Zhao DD, Bao SJ, Zhou WJ, Li HL. Preparation of hexagonal nanoporous nickel hydroxide film and its application for electrochemical capacitor. Electrochemistry communications. 2007 May 1;9(5):869-74.

[60] Fujita T, Chen MW. Characteristic length scale of bicontinuous nanoporous structure by fast fourier transform. Japanese Journal of Applied Physics. 2008 Feb 15;47(2R):1161.

[61] Chen Y, Fitz Gerald J, Chadderton LT, Chaffron L. Nanoporous carbon produced by ball milling. Applied physics letters. 1999 May 10;74(19):2782-4.

[62] Ravikovitch PI, Neimark AV. Characterization of nanoporous materials from adsorption and desorption isotherms. Colloids and Surfaces A: Physicochemical and Engineering Aspects. 2001 Aug 31;187:11-21.

[63] Han J. In: Di Ventra M, Evoy S, Heflin JR eds. *Introduction to Nanoscale Science and Technology*. New York: Springer; 2004.36

[64] Ghosh R. *Protein Bioseparation Using Ultrafiltration: Theory, Applications andNew Developments*. London: Imperial College Press; 2002.

[65] Fissell WH,HumesaHD, Fleischmanb AJ, Roy S.Dialysis and nanotechnology: now, 10 years, or never? *Blood Purif* 2007, 25(1):12-17.38.

[66] Nishizawa M, Menon VP, Martin CR. Metal nanotubule membranes with electrochemically switchable ion-transport selectivity. *Science* 1995, 268:700-702.39

[67] Li Q, Luo G, Feng J, Zhou Q, Zhang L, et al. Amperometric detection of glucose with glucose oxidase absorbed on porous nanocrystalline TiO2 film. *Electroanalysis* 2001, 13(5):413-416.40

[68] Joo S, Park S, Chung TD, Kim HC. Integration of a nanoporous platinum thin film into a microfluidic system for non-enzymatic electrochemical glucose sensing. *Anal Sci* 2007, 23:277-281.

[69] Bayley H, Cremer BS. Stochastic sensors inspired by biology. *Nature* 2001, 413:226-230.

[70] Tsujino I, Ako J, Honda Y, Fitzgera PJ. Drug delivery via nano-, micro and macroporous coronary stent surfaces. *Expert Opin Drug Deliv* 2007, 4(3):287-295.

[71] Desai TA, Sharma S, Walczak RJ, Boiarski A, CohenM, Shapiro J, West T, Melnik K, Cosentino C, Sinha PM, Ferrari M. Nanoporous implants for controlled drug delivery. In: Desai TA, Bhatia S eds. *BioMEMS and Biomedical Nanotechnology Volume III Therapeutic Micro/Nanotechnology*. New York: Springer, 2007

[72] Hu K, Lan D, Li X, Zhang S. Electrochemical DNA biosensor based on nanoporous gold electrode and multifunctional encoded DNA– Au bio bar codes. Analytical chemistry. 2008 Dec 1;80(23):9124-30.

[73] Poh HL, Pumera M. Nanoporous carbon materials for electrochemical sensing. Chemistry–An Asian Journal. 2012 Feb 6;7(2):412-6.

[74] Adiga SP, Curtiss LA, Elam JW, Pellin MJ, Shih CC, Shih CM, Lin SJ,

Su YY, Gittard SD, Zhang J, Narayan RJ. Nanoporous materials for biomedical devices. Jom. 2008 Mar 1;60(3):26-32.

[75] Xu Q, editor. Nanoporous materials: synthesis and applications. CRC press; 2013 Jan 4.

Chapter 3

Biosensors: Design, Development and Applications

Phumlani Tetyana, Poslet Morgan Shumbula and Zikhona Njengele-Tetyana

Abstract

The ability to detect even the slightest physiological change in the human body with high sensitivity and accurately monitor processes that impact human nature and their surroundings has led to an immense improvement in the quality of life. Biosensors continue to play a critical role across a myriad of fields including biomedical diagnosis, monitoring of treatment and disease progression, drug discovery, food control and environmental monitoring. These novel analytical tools are small devices that use a biological recognition system to investigate or detect molecules. This chapter covers the design and development of biosensors, beginning with a brief historical overview. The working principle and important characteristics or attributes of biosensors will also be addressed. Furthermore, the basic types of biosensors and the general applications of these biosensors in various fields will be discussed.

Keywords: bio-receptor, transducer, bio-sensing, analyze

1. Introduction

The importance of monitoring vital processes and parameters in various industries has led to the discovery of small analytical devices known as biosensors. The emergence of these devices has provided solutions to various applications including drug discovery, disease diagnosis, biomedicine, food safety and processing, environmental monitoring, defence, and security [1, 2] as depicted. Biosensors are analytical devices used to investigate the presence of an analyte of interest in a sample. By definition, these are self-sufficient integrated devices that provide qualitative and semi-quantitative analytical data through the use of a biological recognition element that is coupled to a transduction element. The sole purpose of these analytical devices is to rapidly provide accurate and reliable information about an analyte of interest in real time [3–6].

Generally, biosensors are composed of three main components as depicted in **Figure 1**. These include a biological sensing element, physicochemical detector or transducer and a signal processing system [8]. Biological sensing elements are used to interact with the analyte of interest to generate a signal. Sensing elements normally include materials such as tissues, microorganisms, organelles, cell receptors, enzymes, antibodies, and nucleic acids. The signal generated through the interaction of the sensing element and the analyte of interest is then transformed to a measurable and quantifiable electrical signal via the transducer. The signal

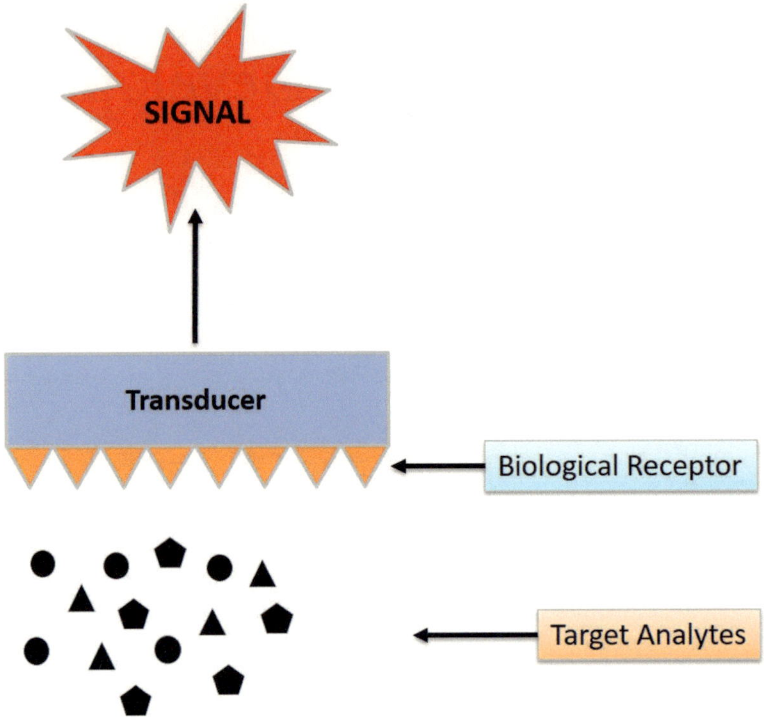

Figure 1.
Basic scheme of a biosensor. Picture adapted from Korotkaya [7] with modifications.

processing system therefore amplifies the electrical signal and conveys it to a data processor that produces a measurable signal in the form of a digital display, print out or color change [9, 10].

The concept of a biosensor is an ancient phenomenon. The first reported concept of a biosensor dates back to 1906 when Cremer [11] discovered that the concentration of an acid suspended in an aqueous solution is equivalent to the electric potential generated between sections of the solution when separated by a glass membrane. This led to the development of the concept of pH by Soren Peder Lauritz Sorensen in 1909, which was followed by the development of an electrode to measure this pH in 1922 by Hughes [12]. This paved way for the development of what is known as a "true biosensor" in 1959 by Leland C. Clark, Jr., who is affectionately known as the "father of biosensors". Clark developed a sensor for detecting glucose in biological samples, using a glucose oxidase electrode that detects the presence of either oxygen or hydrogen peroxide. Since then, great strides have been made in developing highly sensitive and selective biosensing devices [13, 14]. The emphasis of this chapter is on the design, development and applications of biosensors. Various components that constitute a biosensor as well as the working principle of biosensors will be presented. Moreover, various types of biosensors will be highlighted and various fields where these devices are used will also be discussed.

2. Biosensor design

A successful biosensor is composed of two main components, mainly a biological receptor or sensor element and a transducer. A signal processing unit that usually contains a display or printer is normally used in conjunction to a biosensor as depicted in **Figure 2**.

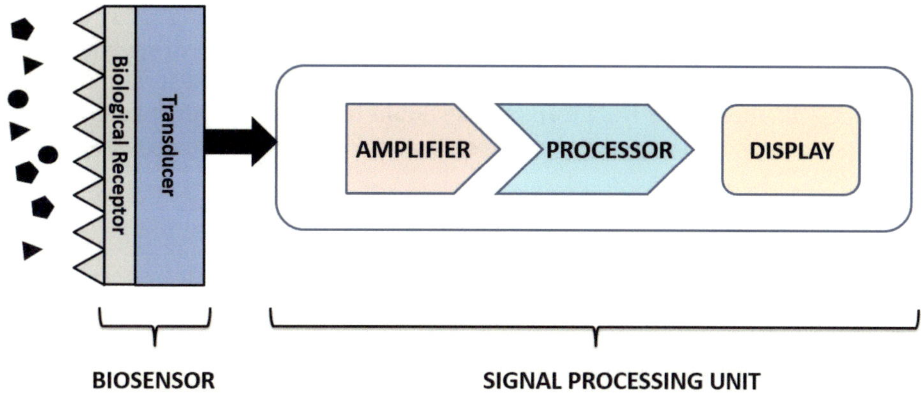

Figure 2.
Biosensor design showing the various components necessary for generating a signal. Picture adapted from [6].

2.1 Biological receptor

This component is also known as a sensor or detector element and is responsible for sensing or detecting the presence and/or the concentration of the target analyte or substance. This is a biological component, which serves as a biochemical receptor that specifically recognizes the target analyte [15]. When the biological receptor interacts with a target analyte, it generates a signal in the form of light, heat, pH, charge or mass change [11]. This material should be highly specific, stable under storage conditions and must be immobilized. Furthermore, the biological receptor should be capable of selectively detecting the target compound or analyte in the test sample. According to Paddle [16], the biological receptor determines the sensitivity of the entire device through the generation of the physicochemical signal that is monitored by the transducer [16, 17].

This component can be a tissue, microorganism, organelle, cell receptor, enzyme, antibody or nucleic acid etc. These can be grouped into two categories, namely catalytic and non-catalytic receptors [18]. The catalytic group of biological receptors are used in devices intended for continuous monitoring of substances at millimolar or micromollar concentrations. These include enzymes, tissues and microorganisms. The non-catalytic group is used mainly in biosensor devices that measure analytes such as steroids, drugs, and toxins etc. which usually occur at very low concentrations (micro to picomollar range). These are non-reusable devices which can only be used once and discarded thereafter. Such receptors include antibodies, antigens, nucleic acids etc. [17, 19, 20].

2.2 Transducer

A transducer forms the second main component in the design of a biosensor. Generally, a transducer is a material that is capable of converting one form of energy to another [11]. In a biosensor, a transducer is responsible for converting the biochemical signal received from the biological receptor, which is a result of the interaction between the target analyte and the biological receptor, into a measurable and quantifiable signal which can be piezo-electrical, optical, electrochemical etc. The transducer detects and measures the change that occurs during biological receptor – analyte interaction [21]. An example of a transducer is a pH sensor in a glucose biosensor. An enzyme, known as glucose oxidase, is used as a biological receptor which binds glucose and converts it to gluconic acid in the presence of oxygen. The pH sensor (transducer) then detects the change in pH (due to production of gluconic acid) and converts it into a voltage change [22, 23]. The following features are recommended when a transducer is designed; specificity to the target analyte, analyte concentration range, response time and suitability for practical applications. Ideally, a transducer should be highly specific to the analyte, give measurement at the lowest analyte concentration within the shortest time possible [24].

3. Working principle of a biosensor

As indicated in the aforementioned sections, a biosensor comprises of a biological receptor coupled with a transducer and signal processing unit, and thus operate on the basis of signal transduction. The combination of these components is designed to convert the biological response into a corresponding electrical response and ultimately a measurable output. In simpler terms, biosensors are responsible for the quantitative analysis of a molecule by relating its biological action into a measurable signal [25]. Initially, the molecule of interest in the test sample binds or interacts specifically with the biological receptor, resulting in a physiological change. This further alters the physicochemical properties of the transducer that is in close proximity to the biological receptor. This further leads to a change in the optical or electronic properties of the transducer which is further converted into an electrical signal which is detectable [26].

The signal generated by the transducer can either be a current or voltage, depending on the type of biological receptor. If the output from the transducer is in the form of a current, then this will be converted into an equivalent voltage. Also, the output voltage is usually very low and masked by a high frequency noise signal, which then requires further alterations, processing and amplification through various filters within the signal processing unit. Finally, the output generated from the signal processing unit should be comparable to the biological quantity being measured [27].

4. Important characteristics of biosensors

Owing to the nature of the applications in which biosensors are used in, several characteristics or parameters have to be met when a biosensor is designed. These characteristics define the performance and usefulness of a biosensor.

4.1 Sensitivity

This is considered as the most important characteristic of a biosensor. The sensitivity of a biosensor is defined as the relationship between the change in analyte concentration and the intensity of the signal generated from the transducer. Ideally, a biosensor should generate a signal in response to small fluctuations in the concentration of the target analyte. Depending on the application, biosensors are required to detect analytes in the ng/ml or fg/ml concentration ranges. This is usually important for medical applications and environmental monitoring purposes [28, 29].

4.2 Selectivity

This refers to the ability of the biosensor to selectively bind and respond only to the desired analyte, in the presence of other molecules or substances. When a signal or response is generated from interactions with an analyte that is different from the target analyte such is termed a false positive result. This is common in biosensors with poor selectivity, thus failing in clinical applications. Selectivity is a very important feature especially in medical applications where the test sample or sample matrix, usually blood or urine, contains numerous molecules that are quite similar to the target analyte and compete for binding to the biological receptor [22, 30].

4.3 Stability

Stability of the biosensor is a very important characteristic especially for biosensors used for continuous monitoring. This feature determines the ability of the biosensor device to resist change in its performance over a period of time in response to interruptions arising from external factors. These can be in the form of temperature, humidity or other environmental conditions. Such interruptions have the potential to induce inaccuracies in the output signal during measurement, thereby affecting the precision and accuracy of the biosensor device [11]. This is because transducers and other electronic components that comprise the biosensor device are mostly temperature sensitive and this can greatly influence their stability. Also, temperature can affect the integrity of the biological receptor as this component tends to degrade with fluctuations in temperature [22].

4.4 Detection limit

A detection limit is defined as the lowest concentration of the target that is able to elicit a measurable signal or response. Ideally, a biosensor should have the lowest detection limit, especially if it is to be used in medical applications where the target analyte might be present at very low concentrations [22].

4.5 Reproducibility

This is also one of the most important features in biosensing, and refers to the ability of the biosensor device to produce matching output signals or results in duplicate experimental runs. The capability of the biosensor to meet this criteria relies on the transducer which is required to perform in a precise and accurate manner [11].

4.6 Response time

This property determines the time it takes for the biosensor to generate a signal or response following the interaction of the biological receptor with the target analyte [26, 27].

4.7 Range or linearity

Biosensor linearity determines the accuracy of the signal obtained, in response to a set of measurements with differing concentrations. This attribute gives insight into the resolution of the biosensor, defined as the minimal change in the target analyte concentration that will elicit a response from the biosensor. This is a very important attribute for a biosensor since most applications require a biosensor to measure a target analyte over wide concentration ranges [11, 22].

5. Considerations for biosensor design

The first step in developing a biosensing device involves investigating the target analyte and understanding how this analyte interacts with certain biological molecules. Once this has been established, the following tasks are critical:

a. Selection of a biological receptor: the specificity and selectivity of a biosensor to the analyte of interest is dependent upon the biological receptor used. A suitable receptor with high affinity for the analyte is thus recommended. Having knowledge of the advantages and disadvantages of various biological receptors in different biosensor applications is very important in selecting a suitable receptor [10, 15, 28].

b. Selection of a suitable immobilization method: for any biological molecule to operate reliably as a biological receptor, it requires attachment onto the surface of a transducer. This process is known as immobilization. Various methods have been used for this task and include adsorption, entrapment, covalent attachment, micro encapsulation and cross linking [31, 32].

c. Selection of a transducer element: the transducer element greatly influences the sensitivity of the biosensor device. Employing the right transducer will result in a device with increased sensitivity while the sensitivity is more likely to be compromised by the use of an ineffective transducer [33, 34].

6. Classification of biosensors

Biosensors are classified according to their biological receptors or transducer elements. **Figure 3** displays a flowchart illustrating the different types of biosensors based on the biological receptors and transducer elements [36]. Some of the biosensors shown in the figure will be discussed further in subsequent sections.

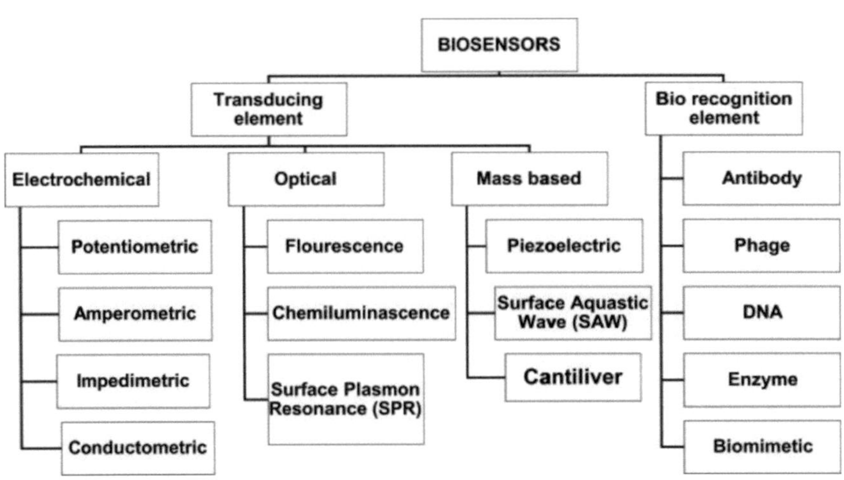

Figure 3.
Flowchart showing the various types of biosensors classified based on their transducing elements and biological recognition elements [35].

6.1 Classification based on biological receptors

6.1.1 Enzyme based biosensors

These type of biosensors form the most researched and reported biosensors based on biological receptors [37, 38]. Enzyme biosensors, useful tools for monitoring rapid changes in metabolite levels in real-time, include pure enzyme preparations or biological processes. They have been derived on immobilization processes such as van der Waals forces, ionic or covalent bonding. In 1967, Updike and Hicks [39] successfully developed a working electrode for the detection of glucose levels and this is considered the first biosensor in the world. The well-known enzymatic biosensors today are glucose and urea biosensors. However, glucose biosensors are most popular among researchers and are reportedly the mostly commercialized biosensors. The glucose biosensor, which was developed by Clark, is made up of glucose oxidase immobilized within a dialysis membrane which is integrated inside oxygen electrodes. Enzymatic biosensors are known for their prolonged use and reusability due to the fact that enzymes used as biological receptors cannot be consumed. Thus, the detection limit and the lifetime of enzyme based biosensors is greatly enhanced by the stability of the enzyme [40].

6.1.2 DNA based biosensors

Another group of biosensors based on a biological receptor is DNA biosensors. The most attractive feature of biosensors is the high selectivity of biosensors for their target analytes in a matrix of chemical or biological elements. DNA biosensors, which use nucleic acids as their biological receptors, detect proteins and non-macromolecular compounds that interact with certain DNA fragments known as DNA probes or DNA primers. The interaction observed stems from the formation of stable hydrogen bonds between the double helix nucleic acid strands [41]. To develop DNA biosensors, immobilization of the probe becomes the most crucial step. The strong pairing of lined up nucleotide strands between bases in their complementary parts influences biosensors based on DNA, RNA, and peptide nucleotide acids to be the most sensitive tool [42]. Lucarelli *et al.* reported that probes, which are short oligonucleotides capable of hybridization with individual areas of the target nucleotide sequence, together with various chemical composition and conformational arrangements, were employed in the development of DNA biosensors. Extremely high sensibility and selectivity is needed to maximize the hybridization efficiency and minimize non-specific binding [43].

6.2 Biosensors based on transduction element

The most commonly applied classification of biosensors is based on the type of transduction element used in the sensor. These biosensors are grouped into three main categories, known as electrochemical biosensors, mass-based biosensors and optical-based biosensors. The working principles of each of the three biosensors are different and can thus be implemented in a variety of applications. Below is a brief description of the different types of biosensors and their working mechanisms. Some of the subclasses under the types of biosensors will also be explained.

6.2.1 Electrochemical biosensors

Electrochemical biosensors, which are the best in the detection of hybridized DNA, DNA binding drugs, glucose concentration, etc., measure the electrical potential difference caused by an interaction between an analyte and the membrane/sensor surface. There is proportionality between the electrical potential difference and the logarithm of the electrochemically active concentration of the material. The current flowing through the system or the potential difference between the electrodes as a result of the redox reactions involving the analyte are employed for its quantification in the sample. Electrochemical biosensors have gained popularity as compared to optical biosensors in the sense that they do not suffer from the many disadvantages optical biosensors experience. They have a more stable output, high sensitivity, fast response and are not prone to interferences. Electrochemical measurements are mostly preferred for sensing applications [44–47]. Electrochemical biosensors can further be classified into various types based on the measuring electrical parameters. These include conductimetric, amperometric, potentiometric and impedimetric sensors [48].

6.2.1.1 Conductometric biosensors

Conductometric biosensors measure the electrical conductivity of the solution in the course of a biochemical reaction. When electrochemical reactions produce ions or electrons, the overall conductivity or resistivity of the solution changes. Due to poor signal-to-noise ratio, they are less commonly used in biosensing applications, particularly when the biological receptor used is an enzyme. However, these biosensors remain useful in the detection of affine interactions [49, 50].

6.2.1.2 Potentiometric biosensors

Potentiometric biosensors measure changes in pH and ion concentrations resulting from antigen/antibody interactions. Although potentiometric biosensors are the least common of all biosensors, different strategies for the development of these biosensors are found. The working principle relies on the fact that when a voltage is applied to an electrode in solution, a current flow occurs because of electrochemical reactions. The voltage at which these reactions occur indicates a particular reaction and particular analyte. Some of the known potentiometric biosensors include those used for the detection of *Neisseria meningitides*, *Brucella melitensis* and *Francisella tularensis* species [51, 52]. Similarly, Hu *et al.* included a light-addressable potentiometric sensor in a microfluidic system to monitor the metabolism of human breast cancer cells in real time [53].

6.2.1.3 Amperometric biosensors

This is perhaps the most common electrochemical detection method used in biosensors. This high sensitivity biosensor can detect electroactive species present in biological test samples [54]. Amperometric-based biosensors detect the difference in current potentials during redox reactions when antigen/antibody pairing occurs. The most common amperometric biosensors use the Clark oxygen electrode. Amperometric biosensors have been developed for the indirect detection of *E. coli* by Nakamura and co-workers [55]. Another amperometric biosensor for the detection of Salmonella Species was developed by Brookes and colleagues [56].

6.2.1.4 Impedimetric biosensors

Impedimetric-based biosensors monitor changes in impedances upon antigen/antibody interaction. Impedance, which usually employs a circuit bridge as a measurement tool, is well suited for detection of bacteria in clinical specimens, to monitor quality and to detect specific food pathogens. Moreover, these biosensors are useful in controlling industrial microbial processes [57].

6.2.2 Mass based biosensors

Piezoelectric biosensors are a group of analytical devices working on a principle of affinity interaction recording. A piezoelectric platform or piezoelectric crystal is a sensor part working on the principle of change in oscillations due to mass bound on the piezoelectric crystal surface. Piezoelectric biosensors, which are considered as mass-based biosensors, produce an electrical signal when a mechanical force is applied. An example of piezoelectric biosensor is the quartz crystal microbalance (QCM) model. The working principle of QCM is depicted in **Figure 4**. Quartz crystal microbalance (QCM) is a very popular tool that is used extensively in the electronic industry. Currently, these tools are used as attenuators in electronic devices and they have a typically fundamental mode frequency of 1–20 MHz. Though higher frequencies provide good opportunities for a sensitive assay, QCM with high frequencies have been reported to exhibit several drawbacks such as their fragility and also the technologically demanding equipment needed for their manufacture [58]. The basic material used in the development of the QCM sensor consists of quartz crystal, which is equipped with metal electrodes. A sensitive coating material on the sensor surface is used to enable detection of the target analyte in the environment. An appropriate electronic circuit is necessary to make conversion of the measured quantity to an electrical signal [59].

6.2.3 Optical biosensor

Optical biosensors are based on the interaction of a sensing element with electromagnetic radiation. They consist of a light source, as well as numerous optical components to generate a light beam with specific characteristics and to beeline this light to a modulating agent, a modified sensing head along with a photodetector. An optical surface plasmon resonance (SPR) biosensor can detect the refractive index changes on the surface of sensor chips, label-free and in real-time. Although different optical methods such as absorption, fluorescence,

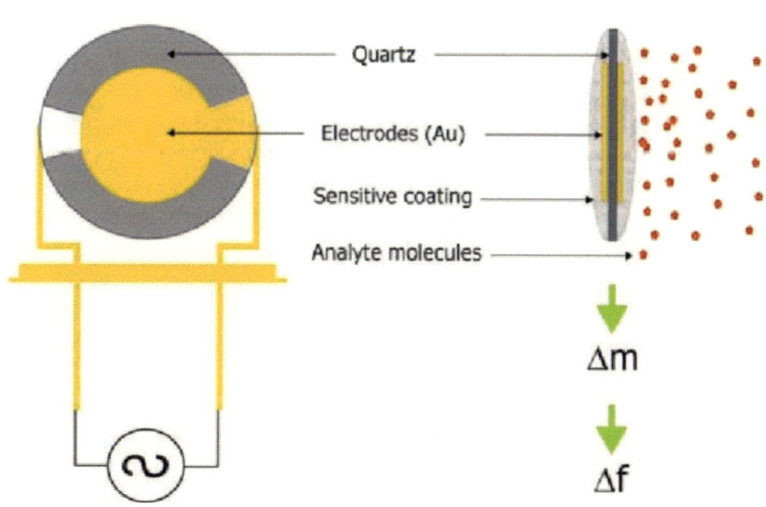

Figure 4.
Basic working principle of Quartz Crystal Microbalance (QCM) sensor [59].

luminescence, internal reflection, surface plasmon resonance, or light scattering spectroscopy utilized herein are becoming popular, fluorescence and surface plasmon resonance enabled spectroscopies still remain the most and widely researched and applied methods [60, 61].

6.2.3.1 Surface plasmon resonance based biosensors

Over the last two decades, surface plasmon resonance (SPR) based biosensors have emerged as important and useful tools due to their unique features for real-time and label-free detection of biomolecular interactions [62, 63]. SPR technology has opened a new avenue for many important applications in the field of sensing due to their attractive sensing capabilities, light weight, compactness and easy implementation [64–67]. The SPR phenomenon has been widely used in biosensing, chemical sensing and environmental sensing applications such as protein–protein hybridization [68, 69], enzyme detection [70, 71] and protein-DNA hybridization. Surface plasmon resonance (SPR), as a physical phenomenon, is not restricted only to events occurring in thin planar metal films. A broad spectrum of differently nanostructured surfaces as well as noble metal nanoparticles are frequently employed for fabrication of SPR-based assays [72–75].

However, conventional commercial SPR-based biosensors and experimental devices are often represented by instruments, which utilize Kretschmann's scheme of plasmon excitation [65]. SPR-based biosensors can be employed to characterize interactions between biomolecules immobilized onto the metal film sensor surface and their counterparts in liquid sample in real time and without labelling. Indeed, these biosensors are actively used to measure binding constants, kinetics of biomolecular interactions and to perform concentration measurements [66]. In turn, these applications make SPR-based biosensors very useful in pharmacological, biomedical, environmental and food studies.

The first practical sensing application of SPR sensors for biomolecular detection was reported by Liedberg and Nylander in 1983 [67]. Since then, SPR biosensors have experienced rapid development in the last two decades and have become a valuable platform for qualitative and quantitative measurements of biomolecular interactions with the advantages of high sensitivity, versatile target molecule selection, and real-time detection. For this reason, SPR sensors are now widely adopted for meeting the needs of biology, food quality and safety analysis, and medical diagnostics.

Over the past decade, many SPR sensors have been reported in applications such as biomolecular interaction analysis, medical diagnostics, environmental monitoring, and food safety [69, 71, 73]. Traditional SPR devices generally require expensive equipment, complicated optics, and precise alignment of the components [74, 75], features that hinder the development of a portable device. Current portable SPR devices still require a portable computer to run the instrument and are about the size of a lunch box.

7. Applications of biosensors

Conventional 'off-site' analysis requires the samples to be sent to a laboratory for testing. These methods allow the highest accuracy of quantification and the lowest detection limits, but are expensive, time consuming and require the use of highly trained personnel. Due to the above drawbacks, there has been a great interest in the technology of biosensors. There has been a phenomenal growth in the field of biosensor development in recent years with emerging applications in a wide range of disciplines. These include environmental monitoring, disease detection, food safety, defence, drug discovery and many more as depicted in **Figure 5** below. A summary of the few and selected representatives and examples of developed applications of biosensors is given below.

7.1 Food industry

Biosensors have been used extensively in the food industry for quality control and assurance purposes. These include applications in the agricultural field during crop production and also during food processing. Quality control remains a major part of food production and is responsible for the production of healthy food with a prolonged shelf life and also complies with regulations. Biosensors have been used as on-line or at-line quality sensors that make it possible for quality sorting, automation and reduction of production cost and production time. Also, biosensors have been developed to detect particular compounds in foods. These devices detect chemicals or biological agents that contaminate food or might indicate the presence of unwanted substances in food. Moreover, biosensors have been developed for monitoring and estimating cross-contamination of surfaces and food products [77–80].

7.2 Environment

Environmental pollution has an impact on human health and can therefore compromise the quality of life. Depending on the purpose, sensitive and selective methods are needed for both quantitative and qualitative determination of target analytes. Biosensors have found widespread use in environmental monitoring for the detection of chemical agents, organic pollutants, potentially toxic elements and pathogens that might pose a health hazard. Biosensors such as immunosensors, aptasensors, genosensors and enzymatic biosensors are amongst the most preferred

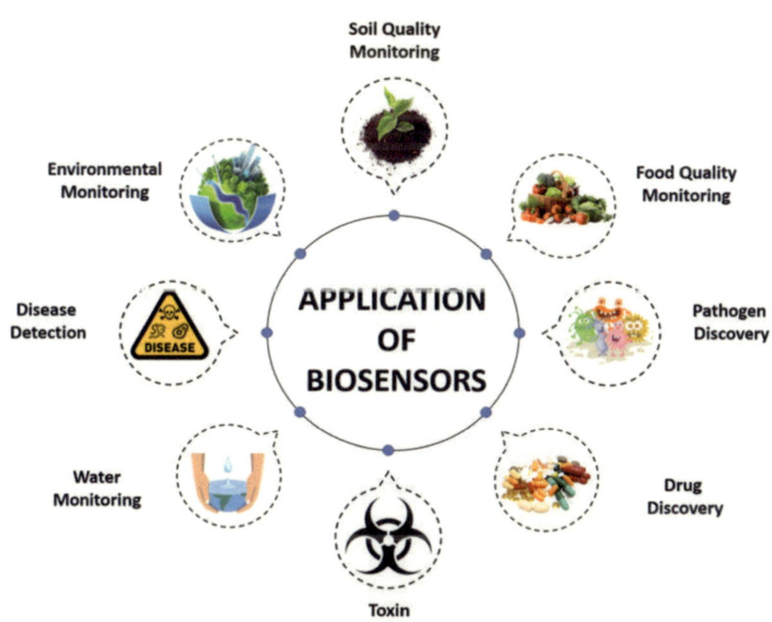

Figure 5.
Various applications where biosensors have been used. Picture adapted from [76].

for environmental monitoring. These are known to use antibodies, aptamers, nucleic acids and enzymes as biological receptors. For example, a biosensor was developed to detect pesticides such as organophosphate and carbamate and also monitor their effects on the environment. Biosensors detect pollutants by measuring colour, light, fluorescence or electric current [81–84].

7.3 Medical

Most of the biosensors reported in the past years are found to be based on the phenomena of molecular interactions which are essentially employed in various forms at different scales. In the discipline of medical science, the applications of biosensors are growing rapidly. Some of the applications that have benefited from the emergence of biosensors include cancer detection and monitoring, cardiovascular disease monitoring, and diabetes control. Cancer diagnosis and treatment are of great interest due to the widespread occurrence of the diseases, high death rate, and recurrence after treatment. In medicine, biosensors can be used to monitor blood glucose levels in diabetics, detect pathogens, and diagnose and monitor cancer progression [85]. The use of emerging biosensor technology could be instrumental in early detection of cancer for effective treatment administration [86]. By measuring levels of certain proteins expressed and/or secreted by tumor cells, biosensors can detect presence of a tumor, whether benign or cancerous, and also give information of whether treatment is effective in reducing or eliminating such cancerous cells [87, 88].

Cardiovascular diseases, which are the primary cause of death are still considered as one of the biggest dilemma the world is facing with about one million people suffering from it. The ability to detect such diseases earlier may result in the reduction of mortality cases. Some of the sensing techniques that have been used herein include immunoaffinity column assay, fluorometric assays, and enzyme-linked immunosorbent assay [89–91]. However, the above techniques are laborious, and therefore require well trained and qualified personnel and are time consuming. Therefore, biosensors are being used for the detection of cardiac markers and early diagnosis. Biosensors have been reported to offer vast advantages over conventional diagnosis assays since they are established on electrical measurements and also employ biochemical molecular recognition elements which gives a desired selectivity with a particular biomarker of interest [92, 93].

8. Conclusions

Biosensors continue to offer solutions and control of various processes across a range of applications. As technology advances, new methods that will result in the development of even better biosensors are emerging, and these seek to address all limitations associated with these devices. The development of biosensors revolves around their sensitivity, specificity, cost effectiveness and ability to detect small molecules. This is mostly determined by the right combination of a biological receptor and a transducer element, components which form the basis of a biosensor.

Acknowledgements

The authors would like to thank the DSI/Mintek Nanotechnology Innovation Centre for financial assistance towards this project.

Conflict of interest

Authors report no conflict of interest.

Author details

Phumlani Tetyana[1*], Poslet Morgan Shumbula[2] and Zikhona Njengele-Tetyana[3]

1 Department of Science and Innovation/Mintek Nanotechnology Innovation Centre, Advanced Materials Division, Mintek, Randburg, Johannesburg, South Africa

2 Department of Chemistry, University of Limpopo, Sovenga, South Africa

3 Centre for Metal Based Drug Discovery, Advanced Materials Division, Mintek, Randburg, Johannesburg, South Africa

*Address all correspondence to: phumlanit@mintek.co.za

References

[1] Monosik R, Stredansky M, Sturdik E. Biosensors — classification, characterization and new trends. Acta Chimica Slovaca. 2012;5(1);109-120, DOI: 10.2478/v10188-012-0017-z

[2] Turner APF. Biosensors: sense and sensibility. Chem. Soc. Rev., 2013; 42; 3184-3196. DOI:10.1039/c3cs35528d

[3] Mehrotra P. Biosensors and their applications – A review. Journal of Oral Biology and Craniofacial Research. 2016;6;153-159. http://dx.doi.org/10.1016/j.jobcr.2015.12.002

[4] Thevenot D.R., Toth K., Durst R.A., Wilson G.S. Electrochemical biosensors: recommended definitions and classification. Pure Appl Chem. 1999;71:2333-2348.

[5] Damiati S, Schuster B. Electrochemical Biosensors Based on S-Layer Proteins. Sensors. 2020; 20; 1721; doi:10.3390/s20061721

[6] Sabr AKH. Biosensors. American Journal of Biomedical Engineering. 2016; 6(6); 170-179. DOI: 10.5923/j.ajbe.20160606.03

[7] Korotkaya EV. Biosensors: Design, Classification, and Applications in the Food Industry. Foods and Raw Materials. 2014;2 (2);161-171. DOI 10.12737/5476

[8] Malhotra S, Verma A, Tyagi N, Kumar V. Biosensors: principle, types and applications. Int. J. Adv. Res. Innov. Ideas Educ. 2017; 3 (2); 3639-3644

[9] Malik P, Katyal V, Malik V, Asatkar A, Inwati G, Mukherjee TK. Nanobiosensors: Concepts and Variations. International Scholarly Research Notices. 2013; 2013; doi.org/10.1155/2013/327435

[10] Grieshaber D, MacKenzie R, Voros J, Reimhult E. Electrochemical Biosensors - Sensor Principles and Architectures. Sensors. 2008; 8; 1400-1458. DOI: 10.3390/s80314000

[11] Bhalla N, Jolly P, Formisano N, Estrela P. Nikhil Bhalla, Pawan Jolly, Nello Formisano and Pedro Estrela. Essays in Biochemistry. 2016;60;1-8. DOI: 10.1042/EBC20150001

[12] Hughes WS. The potential difference between glass and electrolytes in contact with the glass. Journal of American Chemical Society. 1922; 44; 2860-2867. DOI: 10.1021/ja01433a021.

[13] Clark LC, Lyons C. Electrode systems for continuous monitoring cardiovascular surgery. Annuals of the New York Academy of Sciences. 1962; 102; 29-45.

[14] Li Y-CE, Lee IC. The Current Trends of Biosensors in Tissue Engineering. Biosensors. 2020; 10(88); 1-22. doi:10.3390/bios10080088

[15] Chaubey A, Malhotra BD. Mediated biosensors. Biosensors & Bioelectronics. 2002; 17; 441-456. DOI.org/10.1016/S0956-5663 (01)00313-X

[16] Paddle BM. Biosensors for chemical and biological agents of defence interest. Biosensors & Bioelectronics. 1996; 11 (11); 1079-1113. DOI: 0956-5663/961515.00

[17] Lowe CR. Biosensors. Trends in Biotechnology. 1984; 2(3); 59-65. doi.org/10.1016/0167-7799 (84)90011-8

[18] Castillo J, Gáspár S, Leth S, Niculescu M, Mortari A, Bontidean I, Soukharev V, Dorneanu SA, Ryabov AD, Csöregi E. Biosensors for life quality Design, development and applications. Sensors and Actuators B. 2004; 102; 179-194. DOI:10.1016/j.snb.2004.04.084

[19] Arnold MA, Meyerhoff ME. Recent advances in the development and analytical applications of biosensing probes, C R C Critical Reviews in Analytical Chemistry. 1988; 20; 149-196. doi.org/10.1080/00078988808048811

[20] Pearson JE, Gill A, Vadgama P. Analytical aspects of biosensors. Annals of Clinical Biochemistry. 2000; 37; 119-145.

[21] Thevenot DR, Toth K, Durst RA, Wilson GS. Electrochemical biosensors: recommended definitions and classification. Biosensors and Bioelectronics. 2001; 16; 121-131. DOI.org/10.1016/S0956-5663 (01)00115-4

[22] Wilkins E, Atanasov P. Glucose monitoring: state of the art and future possibilities. Medical Engineering and Physics. 1996; 18; 273-288. doi.org/10.1016/1350-4533 (95)00046-1

[23] Ramirez NB, Salgado AM, Valdman B. The evolution and developments of immunosensors for health and environmental monitoring: Problems and perspectives. Brazilian Journal of Chemical Engineering. 2009; 26 (02); 227-249. DOI: 10.1590/S0104-66322009000200001

[24] Sethi RS. Transducer aspects of biosensors. Biosensors & Bioelectronics. 1994; 9; 243-264. doi.org/10.1016/0956-5663 (94)80127-4

[25] Soleymani L, Li F. Mechanistic Challenges and Advantages of Biosensor Miniaturization into the Nanoscale. American Chemical Society Sensors 2017;2 (4); 458-467. doi.org/10.1021/acssensors.7b00069

[26] Njagi JI, Kagwanja SM. The Interface in Biosensing: Improving Selectivity and Sensitivity. In: Helburn R, Vitha MF, editors. Interfaces and interphases in Analytical Chemistry. American Chemical Society; 2011. p. 225-247. DOI: 10.1021/bk-2011-1062.ch011

[27] Ali J, Najeeb J, Ali MA, Aslam MF, Raza A. Biosensors: Their Fundamentals, Designs, Types and Most Recent Impactful Applications: A Review. Journal of Biosensors & Bioelectronics. 2017;8(1); 1-9. DOI: 10.4172/2155-6210.1000235

[28] Saha K, Agasti SS, Kim C, Li X, Rotello VM. Gold nanoparticles in chemical and biological sensing. Chemical Reviews. 2012; 112; 2739-2779. DOI: 10.1021/cr2001178

[29] Wang Y, Knoll W, Dostalek J. Bacterial pathogen surface plasmon resonance biosensor advanced by long range surface plasmons and magnetic nanoparticle assays. Analytical Chemistry. 2012; 84; 8345-8350. doi.org/10.1021/ac301904x

[30] Polatoğlu I, Aydın L, Nevruz BC, Özer S. A Novel Approach for the Optimal Design of a Biosensor. Analytical Letters. 2020; 1428-1445. DOI:10.1080/00032719.2019.1709075

[31] Morales MA, Halpern JM. Guide to Selecting a Biorecognition Element for Biosensors. Bioconjugate Chemistry. 2018; 29(10); 3231-3239. doi:10.1021/acs.bioconjchem.8b00592

[32] Korotkaya EV. Biosensors: Design, Classification, and Applications in the Food Industry. Foods and Raw Materials. 2014;2 (2);161-171. DOI 10.12737/5476

[33] Sassolas A, Blum LJ, Leca-Bouvier BD. Immobilization strategies to develop enzymatic biosensors. Biotechnology Advances. 2012; 30; 489-511. DOI: 10.1016/j.biotechadv.2011.09.003

[34] Morales MA, Halpern JM. Guide to Selecting a Biorecognition Element for Biosensors. Bioconjugate Chemistry.

2018; 29(10); 3231-3239. doi:10.1021/acs.bioconjchem.8b00592.

[35] Najeeb MA, Ahmad Z, Shakoor RA, Mohamed AMA, Kahraman R. A novel classification of prostate specific antigen (PSA) biosensors based on transducing elements. Talanta. 2017; 168; 52-61. DOI: 10.1016/j.talanta.2017.03.022

[36] Mungroo NA, Neethirajan S. Biosensors for the Detection of Antibiotics in Poultry Industry. A Review. Biosensors. 2014; 4, 472-493. DOI:10.3390/bios4040472

[37] Ferri S, Kojima K, Sode K. Review of glucose oxidase and glucose dehydrogenases. Journal of diabetes science and Technology. 2011; 5; 1068-1076. DOI: 10.1177/193229681100500507

[38] Ali SMU, Nur O, Willander M, Danielson B. A fast and sensitive potentiometric glucose microsensor based on glucose oxidase coated ZnO nanowires grown on a thin layer wire. Sensors and Actuators B: Chemical. 2010; 145; 869-874. DOI:10.1016/j.snb.2009.12.072

[39] Updike S, Hicks G. The enzymatic electrode. Nature. 1967; 214; 986-988

[40] Marquette CA. State of the art and research advances in immunoanalytical systems. Biosensors and Bioelectronics. 2005; 21; 1424-1433. DOI: 10.1016/j.bios.2004.09.037

[41] Wang J. DNA biosensors based on peptide nucleic acid (PNA) recognition layers - A review. Biosensors and Bioelectronics. 1998; 13 (7-8); 757-762. DOI: 10.1016/s0956-5663(98)00039-6.

[42] Monošík R, Streďanský M, Šturdík E. Biosensors-classification, characterization and new trends. Acta Chimica Slovaca. 2012; 5; 109-120. DOI: https://doi.org/10.2478/v10188-012-0017-z.

[43] Lucarelli F, Tombelli S, Minnuni M, Marazza G, Mascini M. Electrochemical and piezoelectric DNA biosensors for hybridisation detection. Analytica Chimica Acta. 2008; 609; 139-159. DOI: 10.1016/j.aca.2007.12.035

[44] Koyun A, Ahlatcolu E, Koca Y. Biosensors and their principles. In A Roadmap of Biomedical Engineers and Milestones; Kara, S., Ed.; InTech: Rijeka, Croatia, 2012

[45] Mungroo NA, Neethirajan S. Biosensors for the Detection of Antibiotics in Poultry Industry. A Review. Biosensors. 2014; 4, 472-493. DOI: 10.3390/bios4040472

[46] Lazcka O, Del Campo FJ, Munoz FX. Pathogen detection: A perspective of traditional methods and biosensors. Biosensors and Bioelectronics. 2007; 22; 1205-1217. DOI: 10.1016/j.bios.2006.06.036

[47] Wang J, Rivas G, Cai X, Palecek E, Nielsen P, Shiraishi H, Dontha N, Luo D, Parrado C, Chicharro M, Farias P, Valera FS. DNA electrochemical biosensors for environmental monitoring: a review. Analytical Chimica Acta. 1997; 347; 1-8. DOI: 10.1016/S0003-2670(96)00598-3.

[48] Huet AC, Fodey T, Haughey SA, Weigel S, Elliott C, Delahaut P. Advances in biosensor based analysis for antimicrobial residues in foods. Trends in Analytical Chemistry. 2010; 29; 1281-1294. DOI: 10.1016/j.trac.2010.07.017.

[49] Karyakin AA, Ulasova EA, Vagin MY, Karyakina, EE. Sensor (Sensor), 2002, no. 1, pp. 16-24

[50] Korotkaya EV. Biosensors: design, classification, and applications in the food industry. Foods and raw materials. 2014; 2(2); 161-171. DOI: 10.12737/5476.

[51] Lee WE, Thomson HG, Hall JG, Fulton RE, Wong JP. Rapid

immunofiltration assay of Newcastle disease virus using a silicon sensor. Journal of Immunological Methods. 1993; 166; 123-131. DOI: 10.1016/0022-1759(93)90336-6.

[52] Thompson HG, Lee WE. Rapid immunofiltration assay of Francisella tularensis. Defence Research Establishment Suffield. 1992; 1376: 1-17.

[53] Hu N, Wu C, Ha D, Wang T, Liu Q, Wang P. A novel microphysiometer based on high sensitivity LAPS and micro-fluidic system for cellular metabolism study and rapid drug screening. Biosensors and Bioelectronics. 2013; 40(1); 167-173. DOI: 10.1016/j.bios.2012.07.010.

[54] Reza KD, Azadeh A, Maryam N, Golnaz R, Morteza AA. Biosensors: Functions and Applications. Journal of Biology and Today's World. 2013; 2 (1); 20-23. DOI: 10.15412/J.JBTW.01020105.

[55] Nakamura N, Shigematsu A, Matsunaga T. Electrochemical detection of viable bacteria in urine and antibiotic selection. Biosensors and Bioelectronics. 1991; 6; 575-580. DOI: 10.1016/0956-5663(91)80022-p.

[56] Brooks JL, Mirhabibollahi B, Kroll RG. Experimental enzyme-linked amperometric immunosensors for the detection of Salmonella in foods. Journal of Applied Bacteriology. 1992; 73; 189-196. DOI: 10.1111/j.1365-2672.1992.tb02977.x

[57] Silley P, Forsythe S. Impedance microbiology: a rapid change for microbiologists Journal of Applied Bacteriology. 1996; 80; 233-243. DOI: 10.1111/j.1365-2672.1996.tb03215.x.

[58] Miroslav P. The Piezoelectric Biosensors: Principles and applications. A review. International Journal of Electrochemical Science. 2017; 12; 496-506. DOI: 10.20964/2017.01.44

[59] Yuwono AS, Lammers PS. Odor Pollution in the Environment and the Detection Instrumentation. Agricultural Engineering International: the CIGR Journal of Scientific Research and Development. Invited Overview Paper, 5

[60] Leatherbarrow RJ, Edwards PR. Analysis of molecular recognition using optical biosensors. Current Opinions in Chemical Biology. 1999; 3; 544-547. DOI: 10.1016/s1367-5931(99)00006-x

[61] Bănică, F-G. (2012). What are chemical sensors? In: Chemical sensors and biosensors. Chichester: Wiley, 1-20

[62] Handrigan JP. Rapid Detection of Food-Borne Pathogens. 2010. http://www.johnpaulhandrigan.net/wp-content/uploads/2012/01/John-Paul-Handrigan

[63] Formisano N, Jolly P, Bhalla N, Cromhout M, Flanagan SP, Fogel R, Limson JL, Estrela P. Optimisation of an electrochemical impedance spectroscopyaptasensor by exploiting quartz crystal microbalance with dissipation signals. Sensors and Actuators B. 2015; 220; 369-375. DOI: 10.1016/j.snb.2015.05.049

[64] Miroslav P. The Piezoelectric Biosensors: Principles and applications. A review. International Journal of Electrochemical Science. 2017; 12; 496-506. DOI: 10.20964/2017.01.44

[65] Kretschmann E. Determination of optical constants of metals by excitation of surface plasmons. Zeitschrift für Physik A Hadrons and nuclei. 1971; 241; 313-324

[66] Schasfoort RBM, Tudos AJ. (2008). How to construct an SPR assay? In Hand book of Surface Plasmon Resonance; RSC Publishing: Cambridge, UK, 2008; 3-9

[67] Liedberg B, Nylander C, Lundstrum I. Surface plasmon

resonance for gas detection and biosensing. Sensors and Actuators B. 1983; 4; 299-304. DOI: 10.1016/0250-6874(82)80008-5

[68] Mungroo NA, Neethirajan S. Biosensors for the Detection of Antibiotics in Poultry Industry. A Review. Biosensors. 2014; 4, 472-493. DOI: 10.3390/bios4040472

[69] Inamor K, Kyo M, Nishiya Y, Inoue Y, Sonoda T, Kinoshita E, Koike T, Katayama Y. Detection and quantification of on-chip phosphorylated peptides by surface plasmon resonance imaging techniques using a phosphate capture molecule. Analytical Chemistry. 2005; 77; 3979-3985. DOI: 10.1021/ac050135t

[70] Ali SMU, Nur O, Willander M, Danielson B. A fast and sensitive potentiometric glucose microsensor based on glucose oxidase coated ZnO nanowires grown on a thin layer wire. Sensors and Actuators B: Chemical. 2010; 145; 869-874. DOI:10.1016/j.snb.2009.12.072

[71] Kanoh N, Kyo M, Inamori K, Ando A, Asami A, Nakao A, Osada H. SPR imaging of photo-cross-linked small-molecule arrays on gold. Analytical Chemistry. 2006; 78; 2226-2230. DOI: 10.1021/ac051777j.

[72] Hu N, Wu C, Ha D, Wang T, Liu Q, Wang P. A novel microphysiometer based on high sensitivity LAPS and micro-fluidic system for cellular metabolism study and rapid drug screening. Biosensors and Bioelectronics, 2013; 40 (1); 167-173. DOI: 10.1016/j.bios.2012.07.010

[73] Li Y, Liu X, Lin Z. Recent developments and applications of surface plasmon resonance biosensors for the detection of mycotoxins in foodstuffs. Food Chemistry.132, 1549-1554. DOI: 10.1016/j.foodchem.2011.10.109

[74] Thiel AJ, Frutos AG, Jordan CE, Corn RM, Smith LM. In situ surface plasmon resonance imaging detection of DNA hybridization to oligonucleotide arrays on gold surfaces. Analytical Chemistry. 69, 4948-4956. DOI: 10.1021/ac0010431.

[75] Jordan CE, Corn RM. Surface Plasmon Resonance Imaging Measurements of Electrostatic Biopolymer Adsorption onto Chemically Modified Gold Surfaces. Analytical Chemistry. 1997; 69; 1449-1456. DOI: 10.1021/ac961012z

[76] Singh S, Kumar V, Dhanjal DS, Datta S, Prasad R, Singh J. Biological Biosensors for Monitoring and Diagnosis. In: Singh J, Vyas A, Wang S, Prasad R, editors. Microbial Biotechnology: Basic Research and Applications. Environmental and Microbial Biotechnology. Springer; 2020. p. 317-335.DOI:10.1007/978-981-15-2817-0_14

[77] Manikandan R, Charumathe N, Fariha BA. Applications of biosensors. Bulletin of Scientific Research. 2019; 1(1); 34-40. DOI: 10.34256/bsr1915

[78] Wei N, Xin X, Du J, Li J. A novel hydrogen peroxide biosensor based on the immobilization of hemoglobin on three-dimensionally ordered macroporous (3DOM) gold-nanoparticle-doped titanium dioxide (GTD) film. Biosensors and Bioelectronics. 2011; 26; 3602-3607. DOI: 10.1016/j.bios.2011.02.010

[79] Villalonga R, Díez P, Yáñez-Sedeño P, Pingarrón JM. Wiring horseradish peroxidase on gold nanoparticles-based nanostructured polymeric network for the construction of mediatorless hydrogen peroxide biosensor. Electrochimica Acta, 56, 4672-4677. DOI:10.1016/J.ELECTACTA.2011.02.108

[80] Rana JS, Jindal J, Beniwal V, Chhokar V. Utility Biosensors for

applications in Agriculture – A Review. Journal of American Science. 2010; 6(9); 353-375.

[81] Justino CIL, Duarte AC, Rocha-Santos TAP. Recent Progress in Biosensors for Environmental Monitoring: A Review. Sensors. 2017; 17; 2918-2943. DOI:10.3390/s17122918

[82] Atkinson AL, Haggett BGD. Whole Cell Biosensors for Environmental Monitoring. Sensor Review. 1993; 13(4); 19 - 22. DOI.org/10.1108/eb007917

[83] Nigam VK, Shukla P. Enzyme Based Biosensors for Detection of Environmental Pollutants - A Review. Journal of Microbiology and Biotechnology. 2015; 25(11); 1773-1781. https://doi.org/10.4014/jmb.1504.04010

[84] Tortolini C, Mazzei F. Electrochemical biosensors for environmental monitoring. International Journal of Environment and Health. 2012; 6(2); 93-110. https://doi.org/10.1039/B403975K.

[85] Tothill IE. Biosensors for cancer markers diagnosis. Seminars in Cell & Developmental Biology. 2009; 20; 55-62. DOI: 10.1016/j.semcdb.2009.01.015

[86] Bohunicky B, Mousa SA. Biosensors: the new wave in cancer diagnosis. Nanotechnology, Science and Applications. 2011; 4; 1-10. DOI: 10.2147/NSA.S13465

[87] Bohunicky B, Mousa SA. Biosensors: the new wave in cancer diagnosis. Nanotechnology, Science and Applications. 2011; 4; 1-10. DOI: 10.2147/NSA.S13465

[88] Tothill IE. Biosensors for cancer markers diagnosis. Seminars in Cell & Developmental Biology. 2009; 20; 55-62. DOI: 10.1016/j.semcdb.2009.01.015

[89] Ooi KGJ, Galatowicz G, Towler HMA, Lightman SL, Calder VL. Multiplex cytokine detection versus ELISA for aqueous humor: IL-5, IL-10, and IFN profiles in uveitis. Investigative Ophthalmology and Visual Science. 2006; 47; 272-277. DOI: 10.1167/iovs.05-0790.

[90] Caruso R, Trunfio S, Milazzo F, Campolo J, De Maria R, Colombo T, Parolini M, Cannata A, Russo C, Paino R, Frigerio M, Martinelli L, Parodi O. Early expression of proand anti-inflammatory cytokines in left ventricular assist device recipients with multiple organ failure syndrome. American Society of Artificial Internal Organs. 2010; 56; 313-318

[91] Caruso R, Verde A, Cabiati M, Milazzo F, Boroni C, Del Ry S, Parolini M, Vittori C, Paino R, Martinelli L, Giannessi D, Frigerio M, Parodi O. Association of preoperative interleukin-6 levels with interagency registry for mechanically assisted circulatory support profiles and intensive care unit stay in left ventricular assist device patients. J Heart Lung Transplant. 2012; 31(6); 625-633. DOI: 10.1016/j.healun.2012.02.006

[92] Watson CJ, Ledwidge MT, Phelan D, Collier P, Byrne JC, Dunn MJ, McDonald KM, Baugh JA. Proteomic analysis of coronary sinus serum reveals leucine-rich 2-glycoprotein as a novel biomarker of ventricular dysfunction and heart failure. Circulation Heart Failure. 2011; 4; 188-197. DOI: 10.1161/CIRCHEARTFAILURE.110.952200

[93] Maurer M, Burri S, de Marchi S, Hullin R, Martinelli M, Mohacsi P, Hess OM. Plasma homocysteine and cardiovascular risk in heart failure with and without cardiorenal syndrome. International Journal of Cardiology. 2010; 141; 32-38. DOI: 10.1016/j.ijcard.2008.11.131

Chapter 4

Plasmonic Nanopores: Optofluidic Separation of Nano-Bioparticles via Negative Depletion

Xiangchao Zhu, Ahmet Cicek, Yixiang Li and Ahmet Ali Yanik

Abstract

In this chapter, we review a novel "optofluidic" nanopore device enabling label-free sorting of nano-bioparticles [e.g., exosomes, viruses] based-on size or chemical composition. By employing a broadband objective-free light focusing mechanism through extraordinary light transmission effect, our plasmonic nanopore device eliminates sophisticated instrumentation requirements for precise alignment of optical scattering and fluidic drag forces, a fundamental shortcoming of the conventional optical chromatography techniques. Using concurrent optical gradient and radial fluidic drag forces, it achieves self-collimation of nano-bioparticles with inherently minimized spatial dispersion against the fluidic flow. This scheme enables size-based fractionation through negative depletion and refractive-index based separation of nano-bioparticles from similar size particles that have different chemical composition. Most remarkably, its small (4 μm × 4 μm) footprint facilitates on-chip, multiplexed, high-throughput nano-bioparticle sorting using low-cost incoherent light sources.

Keywords: plasmonic nanopore, optical tweezers, optofluidics, extraordinary light transmission, nano-bioparticle sorting

1. Introduction

Optical chromatography (OC) is an increasingly adapted technique for label-free sorting and analysis of bioparticles including cells, bacteria, fungi [1–4]. It exploits a lightly focused Gaussian laser beam within a microfluidic channel to create opposing optical scattering and fluidic drag forces. One can leverage these controllable forces to realize selective fractionation of bioparticles in a heterogeneous mixture based on size, morphology or chemical composition (i.e., refractive index variation) [5, 6]. OC technique was first implemented in size-based fractionation of inorganic materials such as polystyrene beads. Later, researchers employed this technique for size-based fractionation and sorting of organic particles including human blood constituents including erythrocytes, monocytes, granulocytes, and lymphocytes [2, 5]. Subsequently, differentiation of micronscale bioparticles with subtle differences [4, 7], including those with size differences as small as 70 nm [8], are shown. In addition to size-based separation, OC technique also offers refractive index-based fractionation capability, allowing separation of bioparticles with minuscule differences in chemical composition, such as *Bacillus anthracis* and *Bacillus thuringiensis* [4] and cells with single gene modifications [3]. Most recently, precise separation capability of OC technique is utilized in realization of multi-stage fractionation approaches enabling network-based microfluidic purification [9, 10]. On the other hand, conventional OC technique suffers from powerful laser beam requirements to create sufficiently strong optical scattering forces [2], as well as multiple off-chip and bulky optical multi-axis positioners that are needed to realize well-controlled laser beam profiles precisely aligned against the fluidic flow [5]. Furthermore, separate sets of light sources and optical components are needed for each processing channels, preventing multiplexed high-throughput operation [10].

In a recent publication, we introduced a novel plasmonic nanopore device that eliminates the shortcomings of the conventional OC technique [11]. Here, we review this hybrid Optofluidic PlasmonIC (OPtIC) device merging light focusing and fluidic flow through a tiny (4 μm × 4 μm footprint) plasmonic microlens housing an integrated nanopore channel. Based on a subwavelength-thick ($\sim$ 200 nm) suspended device structure, our optofluidic approach opens the door to practical, scalable, and high-throughput on-chip particle sorting.

2. OPtIC microlens design

In **Figure 1a**, an OPtIC device that consists of a periodic nanohole array (NHA) defined in a suspended multilayer membrane is shown. The mechanically robust membrane consists of a free-standing 100 nm thick silicon nitride (Si_3N_4) substrate coated with 100 nm thick gold (Au) and a 5 nm thick titanium (Ti) adhesion layer. The total thickness of the microlens is $h = t_{Au} + t_{Ti} + t_{Si3N4} = 205$ nm, whereas the lateral dimension of the finite size NHA is 4 μm × 4 μm. Recent research findings suggest that (quasi)periodic arrays of nanoplasmonic apertures behave as micro-convex lenses focusing broadband incoherent light beams to spot sizes comparable to wavelength of light [11–13]. Such tight light focusing capability can be harnessed to realize sufficiently strong optical scattering forces suitable for OC using collimated broadband light sources [14, 15]. In addition, the finite-size plasmonic NHAs can focus light over a broad wavelength range with focusing characteristics dictated by the lateral dimension of the array and nearly insensitive to sub-structural imperfections [13]. In this respect, our plasmonic nanopore device provides a distinct nanofluidic integration capability through small modifications in the NHA design without degrading its light focusing characteristics. The periodic NHA shown in **Figure 1a** consists of $d = 150$ nm diameter openings with a periodicity of $a = 380$ nm. Enhanced light transmission through the periodic nanohole around this center nanoaperture occurs through the extraordinary optical transmission (EOT) effect [16–19]. The enlarged central nanopore with $d_c = 500$ nm exhibits orders of magnitude smaller fluidic resistance with respect to the neighboring nanoholes, enabling efficient nanofluidic flow through it. The OPtIC device uses inlet and outlet fluidic ports that are on the opposite sides of the NHA [20, 21], as depicted in **Figure 1b**. This design facilitates microfluidic access from either side of the membrane [22]. In our simulations, 50 μm distance in between of the inlet and outlet fluidic ports is chosen to provide a clear path for the focused light beam [20].

Figure 1b depicts the cross-sectional view of nanofluidic flow pattern across the OPtIC device calculated using steady-state finite-element method (FEM) simulations (COMSOL Multiphysics). The overall size of the computational domain in **Figure 1b** is 50 μm × 50 μm × 40 μm. The inlet fluidic flow is directed towards the central aperture, where flow velocity is largest along the optical (z) axis in the vicinity of the optical focal point, whereas an almost symmetric behavior is

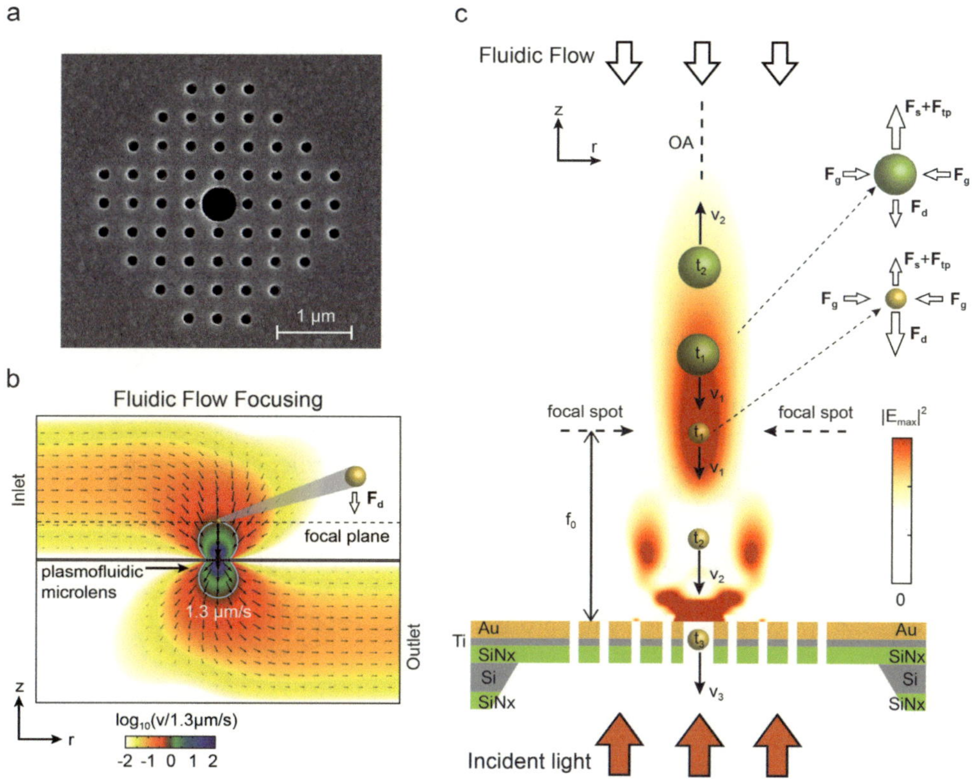

Figure 1.
OPtIC nanopore device enabling selective sorting of bioparticles: (a) top view of OPtIC microlens consisting of a 9 × 9 NHA with enlarged central aperture. (b) Nanofluidic flow pattern across the OPtIC device with 1.3 µm/s flow rate at the focal point. (c) Conceptual illustration of the selective separation mechanism for nano-bioparticles through counter acting forces at the focal point. Copyright 2020 nature publishing group adapted with permission [11].

observed towards the outlet port. The observed fluidic flow pattern can be understood through the Hagen-Poiselle law, where the pressure-driven flow across a cylindrical aperture with hydraulic diameter r_H and thickness h occurs with a volumetric flow rate given by $Q = \Delta p/R_H$ (in $m^3 s^{-1}$). Here, Δp is the pressure gradient across the aperture and $R_H = 8\mu h/\pi r_H{}^4$ (in $Pa{\cdot}s^3 m^{-1}$ with $\mu = 8.9 \times 10^{-4}$ Pa·s being dynamic viscosity of water) is the hydraulic resistance [23]. Since the hydraulic resistance is inversely proportional to the 4th power of hydraulic radius, it is two orders of magnitude smaller across the central aperture with respect to the rest of the holes in the NHA. Hence, convective fluidic flow, which follows the least resistance path, is through the central nanoaperture, as shown in our simulations. In summary, the OPtIC device forces nano-bioparticles to flow towards its focal point, where the dynamic flow trajectories of the particles are aligned with the optical axis of the plasmonic microlens.

A close-up cross-sectional view of the OPtIC device is given in **Figure 1c**, where the fluidic flow is in the *-z* direction within the close vicinity of the focal point (f_0 away from microlens top) as explained above. In this configuration, the collimated light incident from bottom along the *+z* direction is focused by the plasmonic microlens along the optical axis. Here, the focusing pattern, the amplitude-squared electric field ($\sim |E|^2$), is calculated through finite-difference time-domain (FDTD) simulations for incident light at $\lambda = 655$ nm. Forces acting on two different size nano-bioparticles within the focal point region are depicted on the right of the **Figure 1c**. Here, the optical scattering force ($\mathbf{F_s}$) is inherently aligned against the fluidic drag force ($\mathbf{F_d}$) along the optical axis for both particles. In addition, thermo-plasmonic drag force ($\mathbf{F_{tp}}$) caused by electromagnetic heating acts in parallel to $\mathbf{F_s}$. This will be explained in detail later. In addition to radial drag forces ($\mathbf{F_{d,r}}$) due to the fluidic flow, the optical gradient forces ($\mathbf{F_g}$) collimate particles along the optical axis, thus providing a robust mechanism for their precise alignment along the optical axis.

Figure 1c depicts that large particles with diameters above a threshold are driven against the fluid flow (i.e. in the $+z$ direction) provided that $\mathbf{F_s}$ is sufficiently larger than $\mathbf{F_d}$. This is also true for particles with larger refractive indices. Thus, particles with larger diameters and/or higher refractive indices are rejected by the OPtIC device. In contrast, particles with smaller diameters and/or lower refractive indices are propelled through the central nanopore and leave the system from the outlet port. This mechanism provides a complete separation capability for smaller nano-bioparticles (e.g. exosomes) in a heterogeneous mixture through negative depletion. Clogging of the central aperture is prevented by the microlens itself since it keeps the larger particles away from the surface.

Optical radiation force acting on nano-bioparticles can be divided into scattering $\mathbf{F_s}$ and gradient $\mathbf{F_g}$ components, as discussed above, which act along and perpendicular to the optical axis, respectively. While $\mathbf{F_s}$ acts against $\mathbf{F_d}$, $\mathbf{F_g}$ is directed towards the optical axis. Their magnitudes are given by [14]:

$$F_{s,g} = \frac{2n_m P}{c} Q_{s,g} \tag{1}$$

where P is incident light power, n_m is surrounding medium's refractive index, c is the speed of light and $Q_{s,g}$ is a respective dimensionless parameter representing optical pressure transfer efficiency due to reflection/refraction at material interfaces. It can be analytically calculated for simple beam profiles such as lightly focused Gaussian beams acting on a spherical particle. On the other hand, scenarios that use complex beam profiles or target smaller particles with diameters comparable to the optical wavelength (e.g. $d \sim 1$ μm), the ray optics approximation cannot be used. Instead, a Maxwell stress tensor (MST) approach should be adopted [24, 25].

$$T_{ij} = \varepsilon E_i E_j^* + \mu H_i H_j^* - \frac{1}{2}\delta_{ij}\left(\varepsilon|\mathbf{E}|^2 + \mu|\mathbf{H}|^2\right) \tag{2}$$

where $\mathbf{E}$ and $\mathbf{H}$ are electric and magnetic field vectors, ε and μ are the electric permittivity and magnetic permeability of the medium, whereas δ_{ij} is the Kronecker delta symbol. Using MST, the net optical radiation force on a small particle in an arbitrary field profile can be calculated through assuming a bounding box small enough to confine the particle as in [25].

$$F = \oint_S \sum_j \frac{1}{2} \operatorname{Re}\left(T_{ij}\hat{n}_j\right) \tag{3}$$

with S being box surface where $\hat{n}_j$ is a unit vector along one of the principal axes.

3. Focusing efficiency of OPtIC microlens

The central nanopore opening plays a key role in precise alignment of fluidic flow along the optical axis and determining the threshold rejection diameter. On the electromagnetic part, it also controls the focusing characteristics of the OPtIC microlens. Hence, it should be ensured that the focusing behavior does not deteriorate for an admissible d_c range [13]. Focusing patterns of the microlens illuminated by plane waves with λ = 655 nm for various d_c values are given in **Figure 2a**. Compared with the cases of no central nanopore (i.e., d_c = 0 nm) and a uniform NHA (where d_c = 150 nm), the enlarged aperture (d_c = 500 nm) has a negligible effect on light focusing behavior, i.e., maximum intensity ($|E|^2$), spot size and depth of field (DoF). Moreover, focusing behavior does not significantly degrade for a large d_c of 800 nm, as seen on the rightmost panel of **Figure 2a**.

The dashed horizontal lines in **Figure 2a**, corresponding to focal point, indicate that the focal length (f_D) is minimally affected by the change of d_c, where it is 5.32 μm for d_c up to 500 nm, and slightly increases to 5.56 μm for d_c = 800 nm. Intensity profiles along the focal axis (i.e., $z = f_D$, dashed lines in **Figure 2a**) are presented in **Figure 2b**. It is clearly seen that the OPtIC microlens brings incident light to a tight focal spot with a full width at half maximum (FWHM) of 1.12 μm for $d_c \leq 150$ nm, whereas FWHM tends to slightly increase for $d_c \geq 500$ nm, as it becomes 1.24 μm and 1.80 μm for d_c = 500 nm and 800 nm, respectively. Thus, the focusing characteristics of the microlens is remarkably stable for a broad range of d_c. Variation of optical intensity along the optical axis (z-direction) for different d_c is shown in **Figure 2c**, where almost identical behavior is observed for relatively small size central nanopores ($d_c \leq 150$ nm). DoF also shows little variation for d_c up to 500 nm. Only, a slight increase in focal distance is observed when d_c = 800 nm.

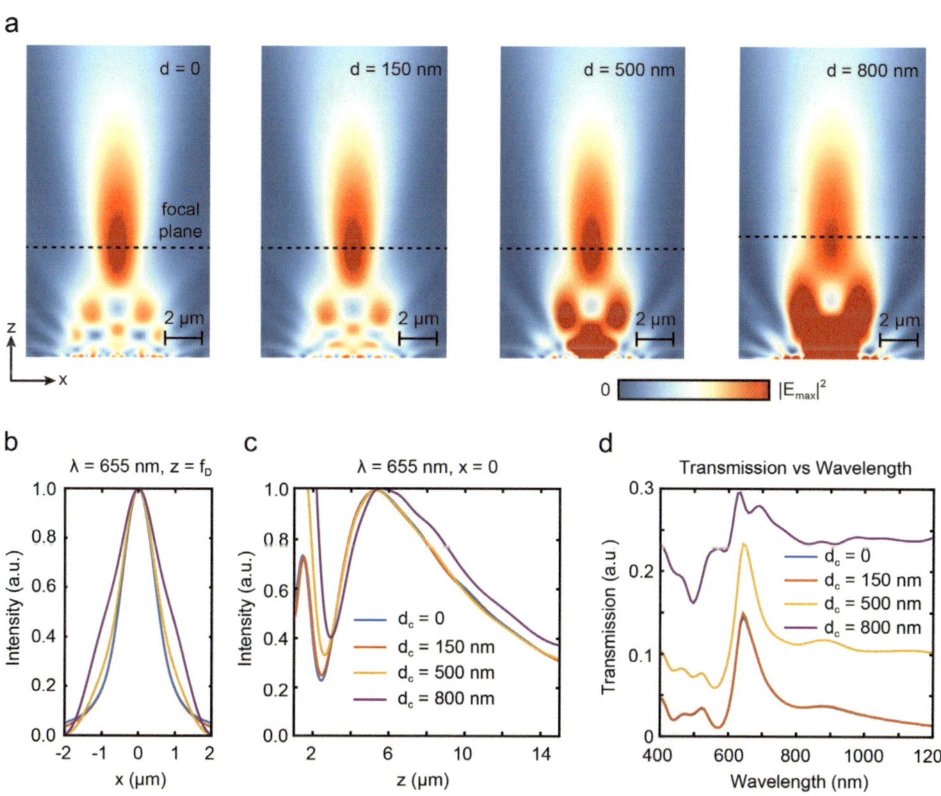

Figure 2.
Monochromatic light focusing behavior with varying central nanopore dimensions: (a) focusing behavior of OPtIC microlens for λ = 655 nm as a function of d_c, corresponding field profile (b) along the focal axis (horizontal dashed lines in (a)) and (c) optical axis. (d) Transmission spectra for plasmonic nanopore devices with different diameter central nanopore openings. Copyright 2020 nature publishing group adapted with permission [11].

The light focusing mechanism of OPtIC microlens relies on the periodic arrangement of smaller nanoholes around the central one. The EOT effect occurs when the Bragg condition is met, i.e., $\mathbf{G} = i\mathbf{G_x} + j\mathbf{G_y}$, where i and j are the corresponding (i, j) grating orders [16, 18]. The transmission spectra shown in **Figure 2d** is obtained using E-field monitors on the focusing side of the microlens. EOT resonance occurs at λ = 650 nm (55 nm FWHM) for the (1,0) grating coupled condition. This confirms that the light focusing behavior is due to interference of in-phase wave components emanating from the periodic NHA [13]. For the NHA with a large central nanopore (d_c = 500 nm), deviation of the EOT peak compared to the NHA without a central opening (d_c = 0 nm) is relatively small, although a larger background transmission is observed. However, further increase in the central nanopore dimensions (d_c = 800 nm) leads to non-resonant light transmission, manifesting itself as enhanced background signal, as shown in **Figure 2d** (top curve). In the light of the above discussion regarding the fluidic flow around the focal point (see **Figure 1**) and light focusing behavior of NHA, an OPtIC microlens with 9 × 9 NHA of d = 150 nm and a = 380 nm, along with a central nanopore opening of d_c = 500 nm is adopted for label-free sorting of nano-bioparticles.

An important observation in **Figure 2a** is the checkerboard-like pattern just above the OPtIC microlens surface, which arises from plasmonic Talbot effect, that is diffractive self-imaging of smaller-diameter nanoholes [13, 26]. On the other hand, intensity right over the central aperture is significantly enhanced for $d_c \geq 500$ nm due to diffractive light transmission through it. These two effects are compared in **Figure 3** using near field phase maps where each small nanohole transmits waves with almost identical amplitude and phase, giving rise to in-phase interference around the focal point. Closer inspection of **Figure 3** reveals that the checkerboard pattern is not disrupted when the central aperture is absent (**Figure 3a**) or is larger than the surrounding nanoholes in the array (**Figure 3c**).

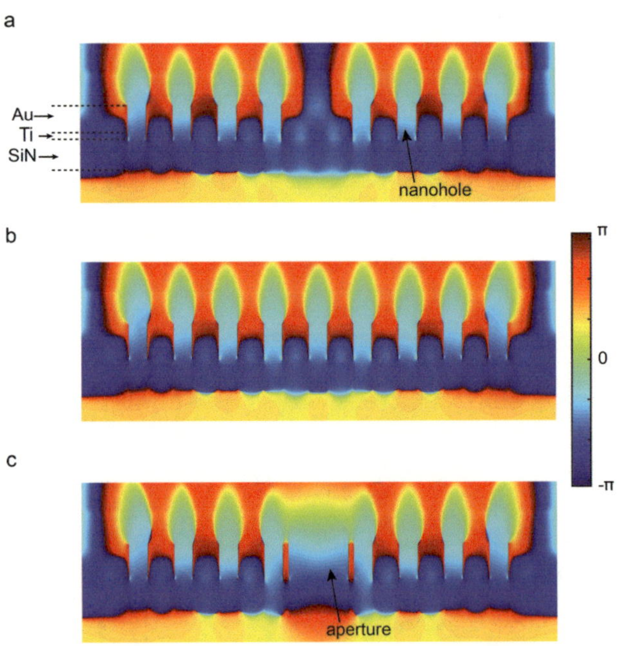

Figure 3.
Light focusing behavior due to in-phase interactions: Near-field phase maps of the hot intensity spots around the OPtIC nanopore device are presented for varying d_c *when illuminated by a monochromatic plane wave* (λ = *655 nm*). d_c *is equal to (a) 0 nm (no central nanopore), (b) 150 nm (identical central opening with the NHA pattern) and (c) 500 nm (enlarged central nanopore). Copyright 2020 nature publishing group adapted with permission [11].*

Light intensity around the central nanopore opening increases with increasing d_c, as seen in **Figure 2a**. This observation is confirmed in **Figure 2c** where a secondary peak in light intensity close to the microlens surface appears. This peak is less intense than the peak around the focal point for smaller size central nanopores ($d_c \leq 150$ nm). However, it is significantly enhanced for larger size nanopore ($d_c \geq 500$ nm). The high intensity region around the nanopore opening may lead to increased optical scattering force $\mathbf{F_s}$, causing undesired rejection of smaller/lower-refractive index particles that managed to pass the focal region and carried towards the central nanopore. However, in addition to tailoring the central nanopore dimension, one can also take advantage of Stokes flow [1, 9, 11], where the fluidic drag forces scale with the relative velocity of nano-bioparticles (**u**) with respect to the flow rate (**v**) of medium (i.e., $\mathbf{F_d} \propto |\mathbf{u}\text{-}\mathbf{v}|$). As fluidic velocity **v** is approximately three orders of magnitude higher close to the central nanopore opening with respect to that of the focal point (**Figure 1b**), fluidic drag forces $\mathbf{F_d}$ are significantly larger too. Hence, optical scattering force $\mathbf{F_s}$ around the nanopore region cannot repel particles that were below the critical diameters. As a result, in the following sections, interplay of forces only around the focal point is considered in the assessment of sorting efficiencies of the OPtIC microlens.

4. Broadband operation of OPtIC microlens

In the preceding section, we showed that the OPtIC nanopore device can focus collimated monochromatic light at λ = 655 nm (close to the EOT resonance peak) into a tight spot. However, broad spectrum operation is also desirable to utilize low-cost broadband incoherent light sources, such as light emitting diodes (LEDs) and halogen bulbs. Hence, development of microlenses that exhibit minimal chromatic aberration is critical [13]. Here, we examine the broadband light focusing behavior of the OPtIC nanopore device with d_c = 500 nm over a wavelength range spanning from 600 nm to 780 nm. **Figure 4a** shows that f_D decreases monotonically with increasing λ with a maximum deviation (Δz_{max}) that is below 200 nm ($\sim\lambda/33$) between 620 nm and 680 nm, which resides within the FWHM of the EOT peak (**Figure 2d**). Below the short-wavelength tail of the EOT peak, Wood's anomaly [19, 27] leads to longer f_D for λ = 600 nm at the transmission minimum (**Figure 2d**).

Optical intensity variation along the focal plane (dashed lines in **Figure 4a**) is shown in **Figure 4b**. Focal point has a FWHM of 1.08 μm, 1.12 μm, 1.24 μm and 1.28 μm at λ = 620 nm, 633 nm, 655 nm and 680 nm, respectively. In addition, the DoF calculated from **Figure 4c** shows small variations within the same spectral window (620 nm < λ < 680 nm). However, the DoF is considerably larger at the wavelength of 600 nm that resides outside the EOT spectral window. Therefore, light focusing characteristics over a sufficiently broad range of wavelengths hinges on the (1,0) resonance transmission (EOT) peak.

Using the Rayleigh-Sommerfeld (R-S) formula [13, 28, 29], the focal length of the finite-size NHA microlens can be calculated:

$$\frac{dI}{dz} = -2I_0 \frac{\pi\rho^2}{\lambda z^2} \sin\left(\frac{\pi\rho^2 n}{\lambda z}\right) = 0 \qquad (4)$$

where I and I_0 are the intensity values calculated at a distance above the lens along the optical (z) axis and the peak intensity, respectively. Here, ρ is the aperture radius and n is the refractive index of the surrounding medium. Comparison of the solid black curve obtained from FDTD simulations to the gray dots calculated via the R-S formula Eq. (4) are in very good agreement for λ between 620 nm and

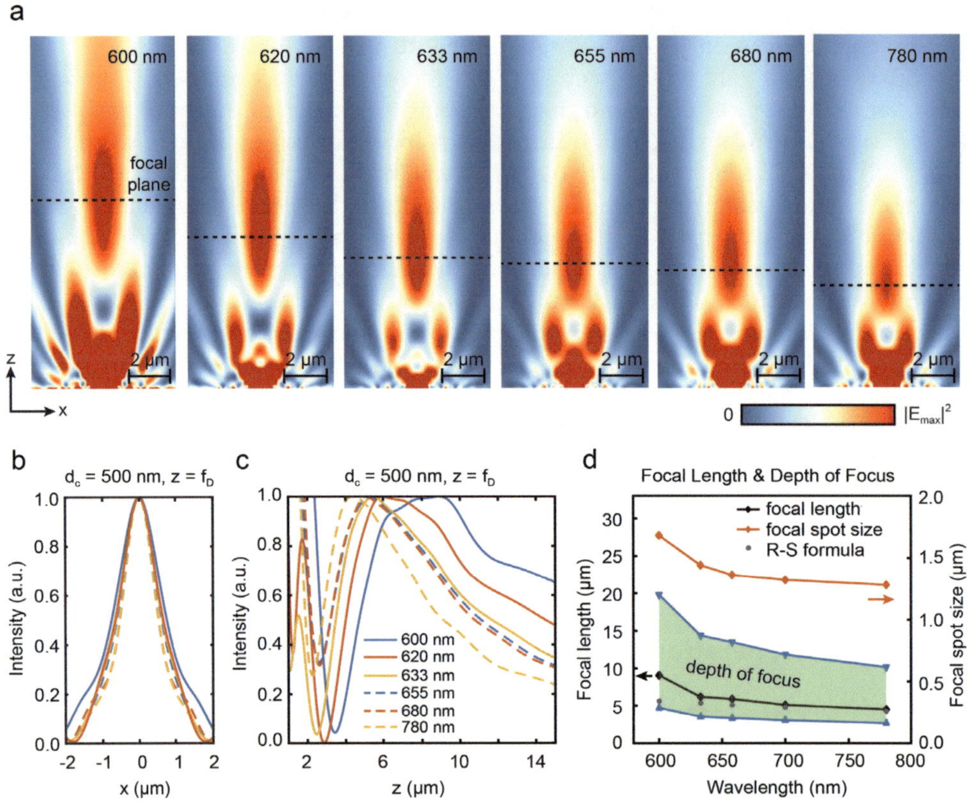

Figure 4.
Broadband light focusing: (a) simulated light focusing behavior of the optimized OPtIC microlens with d_c = *500 nm for a range of* λ, *cross-sectional intensity variation along (b) the focal and (c) optical axis, as well as (d) variation of focusing characteristics in terms of* f_D, *spot size and DoF with* λ. *the gray dots represent values obtained using the Rayleigh-Sommerfeld formula. Copyright 2020 nature publishing group adapted with permission [11].*

680 nm, as seen in **Figure 4d**. The shaded area bounded by the solid blue lines in **Figure 4d** indicates that DoF does not change significantly within the same wavelength range, even though incorporation of an enlarged central aperture breaks the NHA periodicity. Minimal modulation in the focal length and spot size is also shown in **Figure 4d** (the solid black and orange curves, respectively). Thus, the OPtIC microlens exhibits minimal chromatic aberration within the FWHM spread of the EOT peak, offering a well-defined broadband light focusing characteristic.

5. Influence of thermo-plasmonic forces

As conventional wisdom suggests, surface plasmon generation is accompanied by electromagnetic heating, which evokes heat-induced fluid dynamics. The local temperature elevation in the vicinity of the OPtIC microlens induces a buoyance-driven convective fluid flow (Archimedes force) against the reverse main flow stream, resulting in a thermo-plasmonic drag force that drives particles away from the microlens surface [30, 31]. A comprehensive review of this physical mechanism – thermo-induced fluid motion - can be found elsewhere [32, 33]. In this work, thermo-plasmonic drag forces are calculated using Multiphysics FEM simulations, which incorporate electromagnetic (EM) wave, heat transfer and Navier–Stokes equations. Here, we solve the EM wave Equation [34].

$$\nabla \times (\nabla \times \mathbf{E}) - k_0^2 \varepsilon(\mathbf{r})\mathbf{E} = 0 \tag{5}$$

where $k_0 = 2\pi/\lambda_0$ is free-space wavelength and $\varepsilon(\mathbf{r})$ is spatial distribution of ε at λ_0. The calculated **E**-field is used to find the heat source density $q(r) = 0.5\mathrm{Re}[\boldsymbol{J}\cdot\boldsymbol{E}^*]$, which is employed to find the total heat power using $Q = \int q(r)dv$, J being the induced current density in metal [34]. Solving the heat transfer equation simultaneously with the incompressible Navier–Stokes relations,

$$\nabla \cdot \left[\rho c_p T(\mathbf{r})\mathbf{v}(\mathbf{r}) - \kappa \nabla T(\mathbf{r})\right] = Q(\mathbf{r}) \tag{6}$$

$$\rho_0[\mathbf{v}(\mathbf{r}) \cdot \nabla]\mathbf{v}(\mathbf{r}) + \nabla p(\mathbf{r}) - \eta \nabla^2 \mathbf{v}(\mathbf{r}) = \mathbf{F} \tag{7}$$

where ρ, K, c_p and η are the density, thermal conductivity, constant pressure specific heat capacity and dynamic viscosity of the fluidic medium, respectively. $T(r)$ and $\boldsymbol{v}(r)$ are the temperature and the fluidic velocity where $\nabla\cdot \boldsymbol{v} = 0$. Calculations are performed using material parameters that are adopted from the work of Roxworthy et al. [33]. Provided that the EM induced temperature gradient and convective fluid flow distribution are obtained through Eqs. (5)–(7), the volumetric thermo-plasmonic force $\mathbf{F_{tp}}$ and $\mathbf{F_d}$ can be calculated through the Boussinesq approximation [31, 33, 35].

$$\mathbf{F_{tp}} = g\rho_0\beta(T)[T(\mathbf{r}) - T_0]\hat{\mathbf{z}} \tag{8}$$

and Stoke's equation

$$\mathbf{F_d} = -6\pi\eta r\mathbf{v} \tag{9}$$

where g denotes the gravitational acceleration constant and β is the thermal expansion coefficient of the water.

Steady-state 2D temperature spatial distributions in the x-y (**Figure 5a**) and x-z planes (**Figure 5b**) are calculated using Eqs. (5)–(9). To demonstrate the dynamical

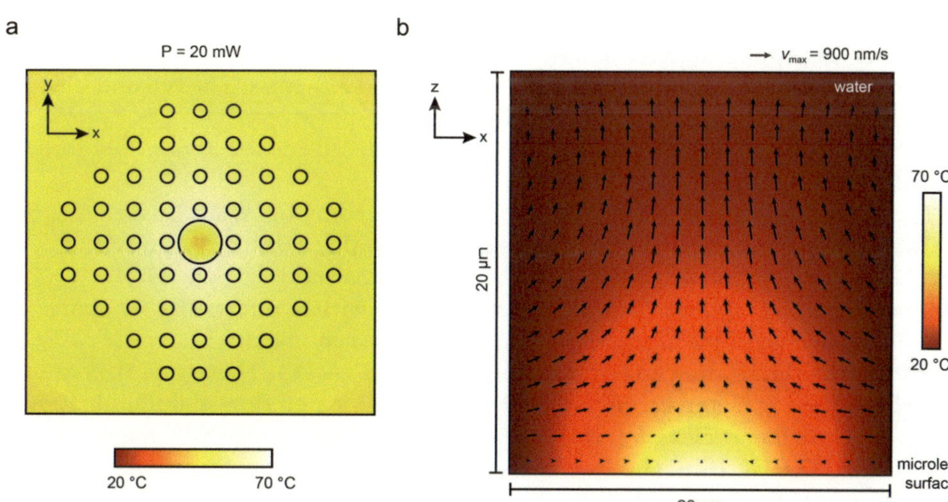

Figure 5.
Thermo-plasmonic heating and Rayleigh–Bénard flow: (a) temperature distribution on the OPtIC nanopore device surface under illumination (633 nm wavelength and 20 mW total power). (b) Temperature distribution and heat-induced fluidic flow pattern on a perpendicular plane crossing the optical axis. The arrows represent the velocity vector $\boldsymbol{v}$*. copyright 2020 nature publishing group adapted with permission [11].*

properties of the thermal-induced fluid convection, the temperature distribution is deliberately overlaid on a vertical 2D slice of convection velocity profile within the same *x-z* plane (**Figure 5b**). In our FEM simulations, the wavelength and power of the excitation light beam is assumed to be 633 nm and 20 mW, respectively. The ambient temperature is T_0 = 20 °C. The ambient temperature near and across the microlens surface is considerably increased due to endogenous heat generation via dissipative losses. Because light transmitting through the central aperture is mostly diffractive in nature and lightly coupled to the surface plasmon polariton (SPP) modes of the NHA, a relatively small temperature increase takes place close to the enlarged aperture region. On the contrary, a significant amount of heat is generated outside the central aperture area, which is attributed to non-radiative damping of SPPs launched on the metal/dielectric interface. Due to the large differences in the heat conductivity of Au and water, heat dissipation occurs slowly along the optical axis of the microlens within the solution environment. This results in reduction of the mass density of ambient water, leading to an upward-directed convective fluidic flow that possesses the features of a toroidal Rayleigh–Bénard flow [33], as represented by the arrows in **Figure 5b**. Here, **v** is along the optical axis in the +*z* direction, exerting a drag force on the suspended nano-bioparticles that directs them away from the microlens surface. Our calculations show that the maximum heat induced convective flow velocity is v_{max} = 900 nm/s, whereas the flow velocity is v = 360 nm/s at the focal point.

6. Label-free sorting of nano-bioparticles

OPtIC nanopore device enables both size- and refractive-index based separation of nano-bioparticles by utilizing a delicate balance of counteracting forces, i.e., $\mathbf{F_s}$, $\mathbf{F_d}$, $\mathbf{F_{tp}}$ and $\boldsymbol{W}$ (gravitational force). Since these forces act along the optical axis within the focal point region, selective particle elution can be readily achieved by adjusting the net force $\mathbf{F_{net}} = \mathbf{F_s} + \mathbf{F_d} + \mathbf{F_{tp}} + \boldsymbol{W}$ [11]. We calculated $\mathbf{F_{net}}$ for varying spherical bioparticles with a mass density of 1.05 g/cm^3 and refractive index of 1.55. The direction of $\mathbf{F_{net}}$ as a function of distance from the microlens ($0 \le z \le 6$ μm) and particle diameter (100 nm $\le d \le$ 1.0 μm) is shown in **Figure 6a** and **b**. In both cases, incident light power is 20 mW and wavelength 633 nm, and the fluidic flow velocities at the focal point $v(z = f_D)$ are 1.3 μm/s and 3.0 μm/s. The red and blue shaded regions in **Figure 6a** and **b** correspond to the positive ($F_s + F_{tp} > F_d + W$, blue shaded region) and negative ($F_s + F_{tp} < F_d + W$, red shaded region) net forces, respectively.

F_{net} is always negative for $d \le 200$ nm, as indicated by the white dashed line on the left in **Figure 6a**. Nano-bioparticles smaller than this threshold diameter, $d < d_{th}$ = 200 nm can travel a path along the optical axis and pass through the nanopore opening towards the outlet port. Particles with diameters larger than d_{th} experience stronger optical scattering and thermo-plasmonic drag forces ($F_s + F_{tp}$) relative to the fluidic drag and gravitational forces ($F_d + W$). Therefore, they are retained in the top chamber. Hence, under the above-stated flow and illumination conditions, small particles ($d \le 200$ nm) get separated from relatively large bioparticles. The OPtIC nanopore device offers selective particle fractionation by adjusting either the incident light power or bulk fluidic flow rate. At a fixed incident light power of 20 mW, when $v(z = f_D)$ is increased to 3.0 μm/s, the threshold particle diameter d_{th} is shifted to 350 nm (**Figure 6b**). By further tuning, relative contributions of acting forces can be tailored for selective separation of nano-bioparticles with diameters up to 500 nm, as larger particles cannot physically pass through the nanopore opening (d_c = 500 nm).

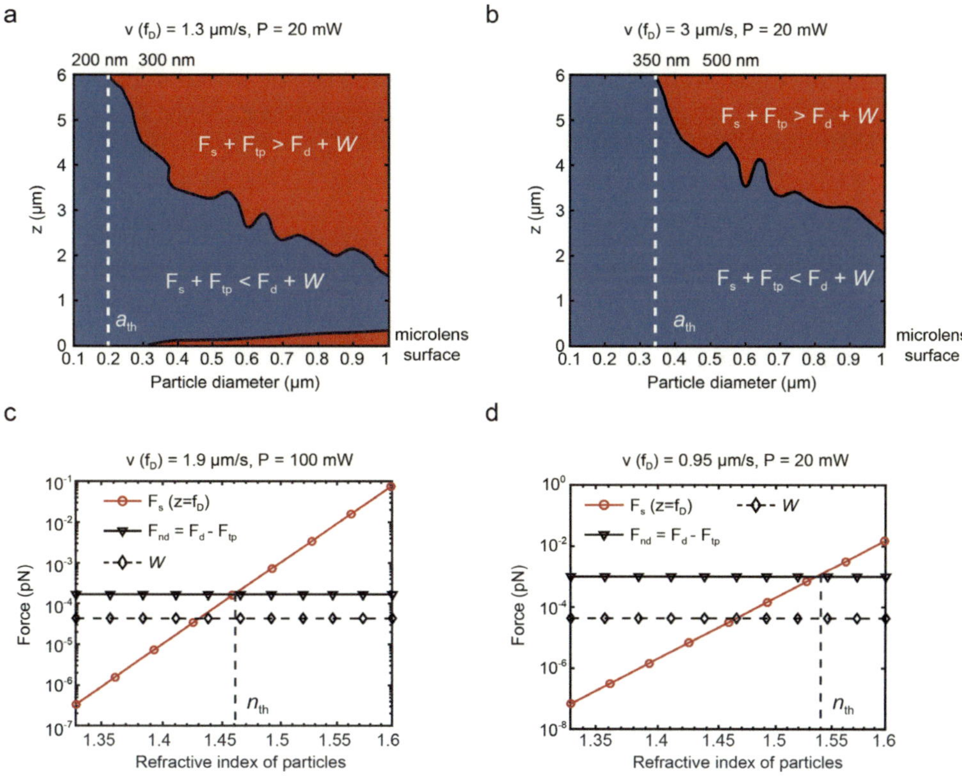

Figure 6.
Label-free selective sorting of nanoparticles: F_{net} *as a function of particle diameter and position along the optical axis is shown for a fluidic flow rate of (a) 1.3 μm/s and (b) 3.0 μm/s at the focal point* $v(z = f_D)$*. A monochromatic light source (633 nm) with 20 mW power is assumed. (c) Forces acting on a fixed diameter (d = 200 nm) nanoparticle at the focal point as a function of particle refractive index* (n_e) *is shown for a fluidic flow velocity of (c)* $v(z = f_D) = 1.9$ *μm/s under 100 mW illumination. (d) Calculations are repeated for a fluidic flow velocity of* $v(z = f_D) = 0.95$ *μm/s under 20 mW illumination. The dashed vertical lines in (a) and (b) represent threshold particle diameters, while those in (c) and (d) denote threshold particle refractive indices* n_e*. The lines in (c) and (d) are first-order polynomial fits to the numerical data. Copyright 2020 nature publishing group adapted with permission [11].*

Size-based separation is not adequate when similar size bioparticles of different origins need to be separated. As indicated in Eq. (1), the optical radiation force acting on particles is a function of both particle radius and refractive index. In this respect, particles comparable dimensions can be separated based on their refractive indices [4, 7]. In general, refractive index is intimately linked to the internal structure and chemical makeup of the nano bioparticles [e.g., exosomes, viruses]. A recent research study has shown that implementation of optical chromatography based on refractive index differences yield successful differentiation of cells with single gene modifications [3]. We calculated forces (F_s, F_{tp}, F_d, W) acting on the nanoparticles (d = 200 nm) as a function of refractive index ($1.33 < n_e < 1.6$) at the focal point ($z = f_D = 5.32$ μm). A fluidic flow rate of $v(f_D) = 1.9$ μm/s and 100 mW monochromatic light illumination at 633 nm is assumed. Solid red and black curves in **Figure 6c** depicts optical scattering F_s and the net drag (i.e., $F_{nd} = F_d - F_{tp}$) forces, respectively. The dashed curves represent contribution of the gravitational force (W). In our calculations, n_e is varied from $n_{water} = 1.33$ to $n_{ps} = 1.6$ (polystyrene beads). As depicted in **Figure 6c**, at the flow rate (1.9 μm/s), the magnitude of F_s increases with n_e and F_s balances F_{nd} when $n_e = 1.46$ (indicated by the vertical dashed line). For smaller n_e, drag force originating from the primary fluidic flow dominates the opposing forces ($F_s + F_{tp}$). Thus, the OPtIC microlens can be employed for selective separation of nano-bioparticles with a threshold of $n_e = 1.46$, corresponding to the refractive index of phospholipids and proteins.

It is known that exosomes (nanovesicles composed mostly of water enclosed by a thin phospholipid membrane) have lower refractive indices n_e in the range of 1.37–1.39 [36–38], which is closer to $n_e = 1.33$ of water. In contrast, virions are tight assemblies of nucleic acids, proteins and lipids with higher n_e around 1.48 [39]. For an exosome-like bioparticle (d = 200 nm) with $n_e \sim 1.38$, F_s can be as small as 4 aN, which is two orders of magnitude smaller than the corresponding value for $n_e = 1.48$ (~400 aN) for a similar size virion. Hence, our OPtIC nanopore device can selectively separate exosomes from similar size virions with high efficiencies.

The threshold n_e can be readily tuned, as in the case of particles size, by adjusting the fluidic flow velocity and/or light power. **Figure 6d** shows that the threshold refractive index can be tuned to $n_e = 1.54$ when the fluidic flow rate is halved to 0.95 μm/s and light power is reduced to 20 mW. Since OPtIC nanopore devices can be operated by directly incident light using an objective-free focusing mechanism, a number of them can be implemented on a planar chip. One can combine multiple OPtIC devices to achieve multi-stage sequential fractionation of nano-bioparticles to realize highly specific size and refractive index-based separation.

7. Radial lining-up of nano-bioparticles

A major limitation of conventional OC is the difficulty of aligning fluidic flow against a lightly focused Gaussian beam in a precise manner [6]. Our OPtIC nanopore device employs a self-collimation mechanism effortlessly aligning fluidic flow along the optical axis of the microlens. This self-collimation capability, lining-up particles against the scattering force, is demonstrated in **Figure 7**. We calculated radial components of the optical gradient ($\boldsymbol{F}_{g\text{-}r}$), fluidic drag ($\boldsymbol{F}_{d\text{-}r}$) and thermo-plasmonic ($\boldsymbol{F}_{tp\text{-}r}$) forces at the focal point $z = f_D$. We analyzed 200 nm (**Figure 7a**) and 600 nm (**Figure 7b**) diameter particles, assuming a fluidic flow velocity of v ($z = f_D$) = 1.3 μm/s at the focal point and a 20 mW monochromatic (633 nm wavelength) light source. As shown in **Figure 7a**, these radial forces act like a

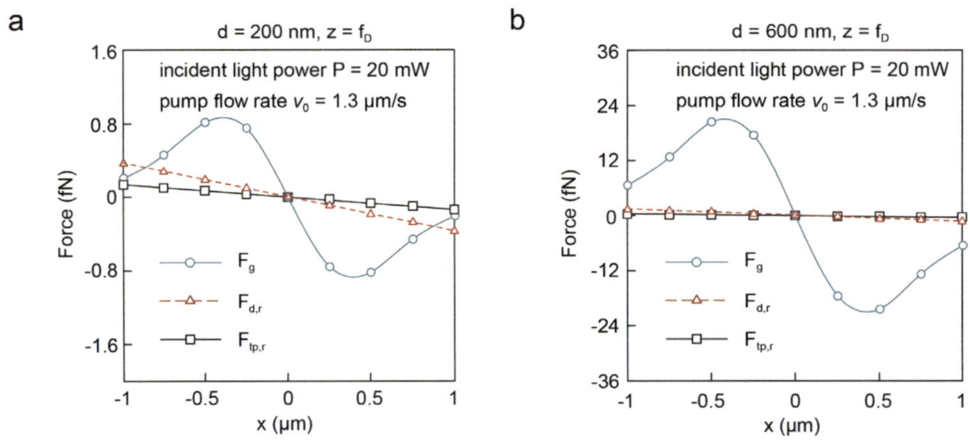

Figure 7.
Self-collimating radial forces along the optical axis: The radial components of $\boldsymbol{F}_g$*,* $\boldsymbol{F}_d$ *and* $\boldsymbol{F}_{tp}$ *in the direction perpendicular to the optical axis is shown for (a) 200 nm and (b) 600 nm particles. The operation conditions are 20 mW light power at 633 nm and a fluidic flow velocity 1.3 μm/s at the focal point. Copyright 2020 nature publishing group adapted with permission [11].*

restoring mechanism ($F \propto -x$) within ± 1 μm away from the optical axis. Nano-bioparticles that deviate from the optical axis are pushed back towards it ($x = 0$) by these radial restoring forces. This leads to minimal spatial dispersion of the nano-bioparticles in the radial direction. As shown in **Figure 7b**, larger particles (d = 600 nm) experience at least an order of magnitude larger optical gradient forces. This enables precise alignment of larger particles against the optical scattering force at the focal point, a critical requirement for selective rejection of particles above a threshold diameter.

As shown in **Figure** 7, nanoparticles in the focal are mainly retained along the optical axis by the optical gradient force $F_{g\text{-}r}$, which is significantly stronger than $F_{d\text{-}r}$. Optical gradient force $F_{g\text{-}r}$ is relatively stable and readily tuned using light intensity. Hence, self-collimation mechanism employed by our OPtIC nanopore device is robust against fluctuations in the fluidic flow rate.

8. Conclusion

In this chapter, we reviewed a facile optofluidic nanopore platform for the purposes of optical chromatography of nano-bioparticles based on their size and/or refractive index (chemical composition). Consisting of a finite-size periodic plasmonic nanohole array on a suspended membrane, this OPtIC nanopore device with an enlarged central aperture facilitates precise alignment of optical scattering, thermo-plasmonic drag, and fluidic drag forces against each other for the purposes of OC. Its self-collimation mechanism eliminates the need for sophisticated and bulky optic components (e.g., lasers, microscope objectives, multi-axis stages, etc.) that are commonly used in conventional OC techniques. Furthermore, our plasmonic microlens opens the door to use of incoherent light source, such as LEDs, for the purposes of OC by readily focusing collimated broadband light into a tight spot. This lensing mechanism provides a robust separation capability that is insensitive to structural variations of the central nanopore. We demonstrated size-based selective sorting of nano-bioparticles, such as exosomes, with a tunable threshold diameter using incident light power or fluidic flow rate. In addition, refractive-index based separation of identical size nano-bioparticles are shown. Similar to size-based separation, the refractive-index (material composition) based separation mechanism is readily tunable through incident light power and fluidic flow rate.

Acknowledgements

A. A. Yanik acknowledges support from National Science Foundation [ECCS-1611290], Gordon and Betty Moore Foundation [GBMF #5263.06], and National Science Foundation CAREER Award [ECCS- 1847733]. X. Zhu was supported by a University of California Chancellor's Dissertation Year Fellowship. We acknowledge Dr. Tom Yuzvinsky for assistance with device fabrication and the W.M. Keck Center for Nanoscale Optofluidics for use of the FEI Quanta 3D.

Conflict of interest

The authors declare no conflict of interest.

Author details

Xiangchao Zhu[1], Ahmet Cicek[2], Yixiang Li[1] and Ahmet Ali Yanik[1,3*]

1 Department of Electrical Engineering, Jack Baskin School of Engineering, University of California Santa Cruz, Santa Cruz, CA, USA

2 Department of Nanoscience and Nanotechnology, Faculty of Arts and Science, Burdur Mehmet Akif Ersoy University, Burdur, Turkey

3 California Institute for Quantitative Biosciences (QB3), University of California Santa Cruz, Santa Cruz, CA, USA

*Address all correspondence to: yanik@ucsc.edu

References

[1] Imasaka T, Kawabata Y, Kaneta T, Ishidzu Y. Optical chromatography. Analytical Chemistry. 1995;67:1763–1765. DOI: 10.1021/ac00107a003

[2] Hebert CG, Terray A, Hart SJ. Toward Label-Free Optical Fractionation of Blood-Optical Force Measurements of Blood Cells. Analytical Chemistry. 2011;83: 5666–5672. DOI: 10.1021/ac200834u

[3] Ma Z, Burg KJL, Wei Y, Yuan X-C, Peng X, Gao BZ. Laser-guidance based detection of cells with single-gene modification. Applied Physics Letters. 2008;92:213902. DOI: 10.1063/1.2938020

[4] Hart SJ, Terray A, Leski TA, Arnold J, Stroud R. Discovery of a Significant Optical Chromatographic Difference between Spores of *Bacillus anthracis* and Its Close Relative, *Bacillus thuringiensis*. Analytical Chemistry. 2006;78:3221–3225. DOI: 10.1021/ac052221z

[5] Kaneta T, Ishidzu Y, Mishima N, Imasaka T. Theory of optical chromatography. Analytical Chemistry. 1997;69:2701–2710. DOI: 10.1021/ac970079z

[6] Makihara J, Kaneta T, Imasaka T. Optical chromatography: Size determination by eluting particles. Talanta. 1999;48:551–557. DOI: 10.1016/S0039-9140(98)00272-0

[7] Hart SJ, Terray AV. Refractive-index-driven separation of colloidal polymer particles using optical chromatography. Applied Physics Letters. 2003;83:5316–5318. DOI: 10.1063/1.1635984

[8] Taylor JD, Terray A, Hart SJ. Analytical particle measurements in an optical microflume. Analytica Chimica Acta. 2010;670:78–83. DOI: 10.1016/j.aca.2010.04.062

[9] Terray A, Hebert CG, Hart SJ. Optical chromatographic sample separation of hydrodynamically focused mixtures. Biomicrofluidics. 2014;8: 064102. DOI: 10.1063/1.4901824

[10] Terray A, Taylor JD, Hart SJ. Cascade optical chromatography for sample fractionation. Biomicrofluidics. 2009;3:044106. DOI: 10.1063/1.3262415

[11] Zhu X, Cicek A, Li Y, Yanik AA. Plasmofluidic microlenses for label-free optical sorting of exosomes. Sci Rep. 2019;9:8593. DOI:10.1038/s41598-019-44801-3

[12] Verslegers L, Catrysse PB, Yu Z, White JS, Barnard ES, Brongersma ML, Fan S. Planar lenses based on nanoscale slit arrays in a metallic film. Nano Letters. 2008;9:235–238. DOI: 10.1021/nl802830y

[13] Gao H, Hyun JK, Lee MH, Yang JC, Lauhon LJ, Odom TW. Broadband plasmonic microlenses based on patches of nanoholes. Nano Letters. 201;10: 4111–4116. DOI: 10.1021/nl1022892

[14] Ashkin A. Acceleration and trapping of particles by radiation pressure. Physical Review Letters. 1970;24:156–159. DOI: 10.1103/PhysRevLett.24.156

[15] Ashkin A, Dziedzic J. Optical trapping and manipulation of viruses and bacteria. Science. 1987;235:1517–1520. DOI: 10.1126/science.3547653

[16] Ebbesen TW, Lezec HJ, Ghaemi HF, Thio T, Wolff PA. Extraordinary optical transmission through sub-wavelength hole arrays. Nature. 1998;391:667–669. DOI: 10.1038/35570

[17] Martin-Moreno L, Garcia-Vidal FJ, Lezec HJ, Pellerin KM, Thio T, Pendry JB, Ebbesen TW. Theory of extraordinary optical transmission through subwavelength hole arrays.

Physical Review Letters. 2001;86:1114–1117. DOI: 10.1103/PhysRevLett.86.1114

[18] Genet C, Ebbesen TW. Light in tiny holes. Nature. 2007;445:39–46. DOI: 10.1038/nature05350

[19] Yanik AA, Wang X, Erramilli S, Hong MK, Altug H. Extraordinary midinfrared transmission of rectangular coaxial nanoaperture arrays. Applied Physics Letters. 2008;93:081104. DOI: 10.1063/1.2973165

[20] Yanik AA, Huang M, Artar A, Chang T-Y, Altug H. Integrated nanoplasmonic-nanofluidic biosensors with targeted delivery of analytes. Applied Physics Letters. 2010;96: 021101. DOI: 10.1063/1.3290633

[21] Huang M, Yanik AA, Chang T-Y, Altug H. Sub-wavelength nanofluidics in photonic crystal sensors. Optics Express. 2009;17:24224–24233. DOI: 10.1364/OE.17.024224

[22] Yanik AA, Cetin AE, Huang M, Artar A, Mousavi SH, Khanikaev A, Connor JH, Shvets G, Altug H. Seeing protein monolayers with naked eye through plasmonic Fano resonances. Proceedings of the National Academy of Sciences of the United States of America. 2011;108:11784–11789. DOI: 10.1073/pnas.1101910108

[23] Oh KW, Lee K, Ahn B, Furlani EP. Design of pressure-driven microfluidic networks using electric circuit analogy. Lab on a Chip. 2012;12:515–545. DOI: 10.1039/C2LC20799K

[24] Okamoto K, Kawata S. Radiation force exerted on subwavelength particles near a nanoaperture. Physical Review Letters. 1999; 83:4534–4537. DOI: 10.1103/PhysRevLett.83.4534

[25] Wang X, Wang X-B, Gascoyne PR. General expressions for dielectrophoretic force and electrorotational torque derived using the Maxwell stress tensor method. Journal of Electrostatics. 1997; 39:277–295. DOI: 10.1016/S0304-3886(97)00126-5

[26] Dennis MR, Zheludev NI, de Abajo FJG. The plasmon Talbot effect. Optics Express. 2007;15:9692–9700. DOI: 10.1364/OE.15.009692

[27] Ghaemi HF, Thio T, Grupp DE, Ebbesen TW, Lezec HJ. Surface plasmons enhance optical transmission through subwavelength holes. Physical Review B. 1998;58:6779–6782. DOI: 10.1103/PhysRevB.58.6779

[28] Ruffieux P, Scharf T, Herzig HP, Völkel R, Weible KJ. On the chromatic aberration of microlenses. Optics Express. 2006;14:4687–4694. DOI: 10.1364/OE.14.004687

[29] Saxena S, Chaudhary RP, Singh A, Awasthi S, Shukla S. Plasmonic Micro Lens for Extraordinary Transmission of Broadband Light. Scientific Reports 2014;4:5586. DOI: 10.1038/srep05586

[30] Baffou G, Girard C, Quidant R. Mapping Heat Origin in Plasmonic Structures. Physical Review Letters. 2010;104:136805. DOI: 10.1103/PhysRevLett.104.136805

[31] Donner JS, Baffou G, McCloskey D, Quidant R. Plasmon-Assisted Optofluidics. ACS Nano. 2011;5:5457–5462. DOI: 10.1021/nn200590u

[32] Kim J. Joining plasmonics with microfluidics: from convenience to inevitability. Lab on a Chip. 2012;12: 3611–3623. DOI: 10.1039/C2LC40498B

[33] Roxworthy BJ, Bhuiya AM, Vanka SP, Toussaint Jr KC. Understanding and controlling plasmon-induced convection. Nature Communications. 2014;5:3173, DOI: 10.1038/ncomms4173

[34] Jackson JD. Classical Electrodynamics. 3rd ed. New York:

John Wiley and Sons; 1999. 832 pp. ISBN: 978–0–471-30932-1

[35] Ndukaife JC, Kildishev AV, Agwu Nnanna AG, Shalaev VM, Wereley ST, Boltasseva A. Long-range and rapid transport of individual nano-objects by a hybrid electrothermoplasmonic nanotweezer. Nature Nanotechnology. 2016;11:53–59. DOI:10.1038/nnano.2015.248

[36] van der Pol E, de Rond L, Coumans FAW, Gool EL, Böing AN, Sturk A, Nieuwland R, van Leeuwen TG. Absolute sizing and label-free identification of extracellular vesicles by flow cytometry. Nanomedicine. 2018;14:801–810. DOI: 10.1016/j.nano.2017.12.012

[37] van der Pol E, Coumans F, Varga Z, Krumrey M, Nieuwland R. Innovation in detection of microparticles and exosomes. Journal of Thrombosis and Haemostasis. 2013;11:36–45. DOI: 10.1111/jth.12254

[38] Gardiner C, Shaw M, Hole P, Smith J, Tannetta D, Redman CW, Sargent IL. Measurement of refractive index by nanoparticle tracking analysis reveals heterogeneity in extracellular vesicles. Journal of Extracellular Vesicles. 2014;3:25361, DOI: 10.3402/jev.v3.25361

[39] Wang S, Shan X, Patel U, Huang X, Lu J, Li J, Tao N. Label-free imaging, detection, and mass measurement of single viruses by surface plasmon resonance. Proceedings of the National Academy of Sciences of the United States of America. 2010;107:16028–16032. DOI: 10.1073/pnas.1005264107

Chapter 5

Adsorption of Heavy Metals from Industrial Wastewater Using Nanoparticles from Agro Wastes

Lekan Taofeek Popoola and Alhaji Shehu Grema

Abstract

Effluents from essential industries have been characterized with heavy metals which are non-biodegradable in nature and also detrimental to health when accumulated in body tissues over long exposure. Adsorption was proved as the best efficient process amongst others to remove these heavy metals from industrial wastewater due to its excellent features. Activated carbons from nanoparticles of agricultural wastes such as pods, shells, husks, peels, shafts and many prepared via calcination process at high temperature can be used as active adsorbent for the industrial wastewater treatment involving heavy metals removal. This chapter discusses heavy metals in industrial wastewater effluents and potential agro wastes from which nanoparticles of activated carbon for industrial wastewater purification could be generated. The transformation of agro wastes nanoparticles into activated carbons via calcination and their applications for heavy metals removal from industrial wastewater via adsorption were examined. Various characterization techniques to study the effects of calcination on structural, morphological and textural properties of activated carbon prepared from agro waste nanoparticles were also discussed. Various isotherm, kinetics, mechanistic and thermodynamics models to investigate the adsorptive nature of the process were presented. Error functions and algorithms for both the linear and non-linear isotherm models regression to affirm their fitness for prediction were presented. Lastly, proposed adsorption mechanisms of heavy metals removal from industrial wastewater using activated carbons from nanoparticles of agro wastes were presented.

Keywords: Adsorption, Industrial wastewater, Nanoparticles, Agro wastes, Heavy metals, Calcination

1. Introduction

Water has been identified as the most essential element needed by man for existence on Earth. However, the natural nature of water is altered when contaminated by waste substances generated from human activities. Water, whose biological, chemical or physical properties have been altered due to the introduction of certain substances, is called wastewater. This renders it useless and makes it unsuitable for industrial and domestic purposes. Some of these contaminants are heavy metals which exist in aqueous waste streams of several industries such as mining operations, dye industries, metal plating, tanneries, alloy industries, smelting, textile industry, radiator manufacturing, and many more. Heavy metals associated with these essential industries include chromium, lead, nickel, cadmium, mercury and iron. They are known with non-biodegradable attribute and their accumulation in body tissues may lead to disruption of functioning cells, kidney congestion, liver damage, nausea, chronic asthma, diarrhea, dermatitis and many more [1, 2]. In order to keep our environment safe from these carcinogenic substances, wastewater treatment techniques are required. Various means of removing heavy metals from wastewater such as coagulation, flocculation, flotation, ion exchange, membrane filtration, electrochemical treatment and chemical precipitation have been presented [3]. However, adsorption was proved to be the most efficient unit operation for the removal of hazardous heavy metals from wastewater solution [4]. Amongst its outstanding features are design and operation simplicity, toxic pollutants insensitiveness, cost effectiveness, high heavy metals sorption efficiency and negligible sludge formation [5]. Nonetheless, nanoparticles synthesized from agro wastes via calcination at higher temperatures have proved to be effectively used as activated carbon for heavy metals removal from industrial wastewater. This chapter general discusses generation of activated carbons from nanoparticles of agro waste as adsorbents for the removal of heavy metals that are usually found in industrial wastewater.

2. Heavy metals in industrial wastewater

The high rate of global industrialization has contributed adversely to the industrial wastewater generated annually. Industrial wastewater contributed to tens of millions of cubic meters per day of the total wastewater discharged globally. In developing countries, record shows that almost 80–90% of all the wastewater is directly discharged untreated into surface water bodies [6]. This causes pollution of water bodies and rivers which consequently endangers the living species around this area and also makes the water unsuitable for usage. Many industrial activities such as petrochemical, electronic, chemical, electrochemical and food processing industries require the use of water for production and cleaning activities. During this process, the water is contaminated by dissolved substances which are needed to be removed before the water is discarded. Industrial wastewater characteristics, quality and volume vary from industry to industry based on the materials they are handling. Industrial wastewater effluent can contain metals, toxic organic compounds, biodegradable organic matter, suspended solids, phosphorus, nitrogen and pathogenic organisms [4]. However, industrial wastewater pollutants of major concern in this context are the heavy metals because of their high toxicity to human beings and plants. They pose risks to human health and the environment at certain concentrations and exposure period. Examples include lead, cadmium, iron, mercury, nickel, zinc, chromium, arsenic, aluminum and barium. **Figure 1** shows the distribution of most widely studied heavy metal ions.

3. Potential agro wastes for industrial wastewater purification

During agricultural practices, some fractions of agricultural biomass which are non-edible in nature are discarded as wastes. These are termed agro wastes and constitute environmental pollution after being discarded. Studies have shown that approximately one-third of 1.6 to 6 billion tons of crops produced globally per year constitutes environmental pollution [8]. However, these agro wastes have exhibited excellent adsorption attributes in removing different pollutants (such as heavy

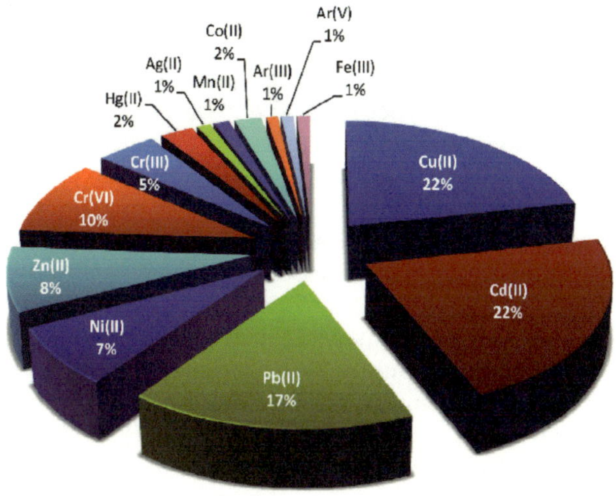

Figure 1.
Distribution of most widely studied heavy metal ions [7].

Figure 2.
Agro wastes for nanoparticles preparation [13].

metals) from wastewater either after chemical or physical modifications; or in their natural state [9]. This could be attributed to active functional groups such as carbonyl, acetamido, phenolic, alcoholic, amino and sulfhydryl which are present in them [10]. Nevertheless, the structural composition of agro wastes to include hemicellulose, lignin, lipids, simple sugars, starch and proteins helps in their adsorption capacity of heavy metals removal from industrial wastewater [11]. Nanoparticles for heavy metals removal from different forms of contaminated water had been prepared from different agro waste such as cotton seed hulls, snail shell, fruit peels, cotton plant wastes, hen feathers, peanut hull, jatropa deoiled cakes, soybean hulls, coconut shells, walnut shells, maize bran, mangosteen shell, wheat shell, *Ceiba pentandra* hulls and many more [12]. **Figure 2** shows different agro wastes that can be transformed into nanoparticles for heavy metals removal from industrial wastewater.

4. Transformation of agro wastes into activated carbon nanoparticles

Nanoparticles (NPs) are broad group of materials that include particulate substances, with dimensions ranging between 1 and 100 nm [14]. They found wide industrial applications when researchers realized the significance of size on the physiochemical properties of a substance [15]. NPs are made up of three layers namely: surface layer (functionalized with a mixture of small metal ions, molecules, polymers and surfactants), shell layer (chemically different material from the core in all aspects), and lastly the core layer (essentially the NP central portion of the NP) [16]. Consequently, studies have shown nanoparticles prepared from agro wastes via physicochemical transformation (such as calcination) as good adsorbents for heavy metals removal from solution [17]. Calcination involves subjecting solids into heating under controlled temperature and environment, purposely to remove volatile substances in order to improve their purity level [18]. In most cases, it is carried out in furnaces which are purposely designed to exclude air such that an inert gas may be replaced. The influence of this method on particle shape, size and crystalline nature of the prepared nanoparticles from agro wastes supports its' suitability for application in industrial wastewater treatment [19]. Factors that determine the physical and chemical properties of activated carbon nanoparticles prepared via calcination of solids (such as agro wastes) include temperature, heating rate, calcination time, particle size and material nature [20].

5. Characterization of nanoparticles from agro wastes

The characterization of nanoparticles prepared from agro wastes for heavy metals removal from industrial wastewater can be categorized into structural; morphological; optical; and particle size and surface area characterizations.

5.1 Structural characterization

The structural characterization gives information about bulk properties, composition and bonding nature of the nanoparticles. These include X-ray diffraction (XRD), energy dispersive X-ray (EDX) and X-ray photoelectron spectroscopy (XPS). However, the first-two are the most prominent. XRD gives detailed information about nanoparticles phase and crystallinity [21]. Information about the crystal lattice is obtained by X-rays, high-energy electrons and neutrons of absorbed or scattered radiation striking the nanoparticles. The diffraction pattern is obtained via elastic interaction between incident radiation of smaller wavelengths and regular arrays of atoms in a crystal lattice. Nanoparticles crystalline structure depends on diffraction angles and intensities in diffracted beams [22]. The atomic numbers of the constituent atoms and their geometrical relationship with respect to the lattice points determine the diffracted angle. The mean crystal size of the nanoparticles can be determined from line widths which could be calculated using Eq. (1) [23].

$$\beta = \frac{K\lambda}{D\cos\theta} \tag{1}$$

where β = line width at half the maximum peak intensity, K = correction factor for nanoparticle shape (taken as 0.9 for spheres), D = crystallite size, and θ is the angle of incidence for the selected diffraction peak.

Another structural characterization usually used in obtaining information about the composition of nanoparticles from agro wastes for treatment of industrial wastewater via X-ray spectrum analysis is the energy dispersive X-ray (EDX) [24]. This is often used in conjunction with scanning electron microscopy (SEM) to obtain nanoparticles elemental composition measured in wt %. The mechanism involves emission of an electron from an inner electron shell of a sample resulting from an incident electron or photon (such as γ –ray or X-ray) hitting an atom at a ground state which leaves a vacancy site in the shell. This electron vacancy site is filled by a more energetic valence electron from the outer shell causing the release of excess energy in the form of an X-ray emission. The intensity of the X-ray emitted by electron beam irradiation is a function of the constituent elements in the nanoparticle. A directly proportional relation exists between the intensity of specific X-ray and the concentration of the explicit element present in the agro waste nanoparticle. **Figure 3** is a typical EDX result obtained for activated carbon prepared from nanoparticles of composite snail shell and rice husk.

The X-ray photoelectron spectroscopy (XPS) is used to know the elements bonding nature and their respective ratio in agro waste nanoparticles. Under this, X-ray photons between 200 and 2000 eV interact with a sample under vacuum via exchange of enough energy with core-level electrons close to the material surface to cause electron ejection. The atomic electron binding energy (E_b) is determined using Eq. (2). A typical XPS spectrum indicating a plot of number of electrons against the electrons binding energy (eV) is obtained for interpretation. Specific set of XPS peaks for each element at its own fingerprint binding energy value can be generated [26]. This characterization technique is highly sensitive and can be used to obtain overall composition and compositional variation of nanoparticles with depth.

$$E_k = hv - E_b - \Phi \quad (2)$$

where E_k = kinetic energy of ejected core electrons, hv = exciting photon energy and Φ = material characteristic work function.

5.2 Morphological characterization

This characterization category is imperative as it gives information about nanoparticles morphology which is an influential factor in determining most of their properties. The most important techniques under this are scanning electron microscopy (SEM) and transmission electron microscope (TEM).

In recent times, morphology of nanoparticles from agro wastes has been analyzed using SEM. It utilizes electrons with small wavelengths which are thermally

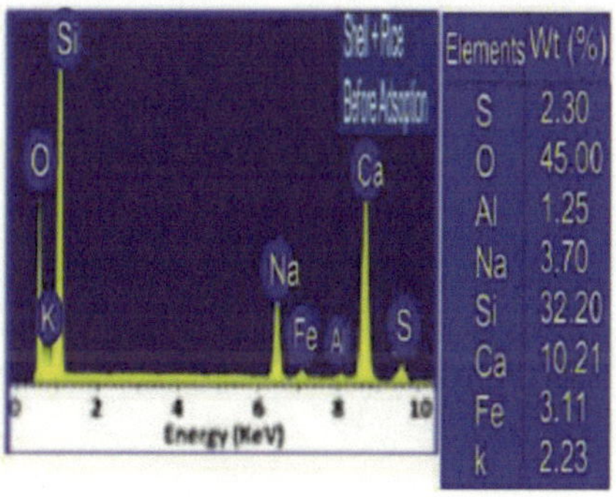

Figure 3.
EDX result of activated carbon from nanoparticles of composite snail shell and rice husk [25].

excited to obtain high resolution of an electron microscope. The electron beam is focused on the nanoparticle sample surface with the aid of electromagnetic lenses to obtain a three-dimensional image of varying magnification and field depth. **Figure 4** is a typical SEM result of activated carbon from nanoparticles of composite snail shell and rice husk.

In a similar manner, TEM is also based on the principle of electron transmittance. However, carbon coated copper grid is used to support the nanoparticle sample during its interaction with the electron beam. The three major parts of TEM include the illumination system where emitted electron beam is allowed to pass through the nanoparticle sample; objective lens and stage; and the imaging system. The image is produced with the aid of connecting devices, intermediate lenses and a projector lens. The image can be taken as either diffraction pattern mode or image mode based on the selection of either the image plane or backfocal plane as the objective plane of the intermediate lens and projector lens.

5.3 Particle size and surface area characterization

The previously discussed XRD, SEM and TEM can be used to determine nanoparticles size and surface area [27]. However, the most common technique for measuring nanoparticles surface area and size is the Brunauer–Emmett–Teller (BET) method [28]. The technique of adsorption and desorption principle is adopted using nitrogen gas. The specific surface area of agro waste nanoparticle is calculated using a monolayer adsorbate weight (W_m) obtained from the slope and intercept of a linear plot of $\frac{1}{W[(P_o/P)-1]}$ against $\left(\frac{P}{P_o}\right)$ in Eq. (3).

$$\frac{1}{W[(P_o/P)-1]} = \frac{1}{W_m C} + \frac{C-1}{W_m C}\left(\frac{P}{P_o}\right) \tag{3}$$

where W = weight of the gas adsorbed at a relative pressure P/P_o, P/P_o = true equilibrium pressure/equilibrium pressure if no gas were adsorbed and C = BET constant which is a function of adsorption energy of the first adsorbed layer. **Table 1** gives information about surface area, total pore volume and average pore diameter of activated carbon prepared from nanoparticles of agro wastes for heavy metals removal from solution.

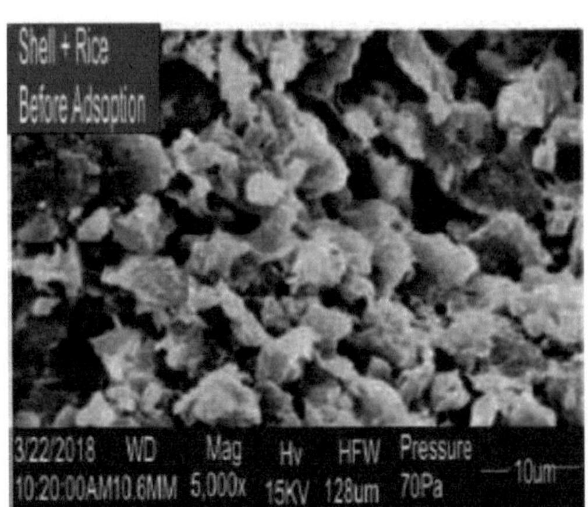

Figure 4.
SEM image of activated carbon from nanoparticles of composite snail shell and rice husk [25].

Agro waste sources for activated carbon	Surface area (m^2/g)	Total pore volume (cm^2/g)	Average pore diameter (Å)	Reference
oil palm frond	700.00	0.3200	5.85	[29]
kenaf core fiber	4.00	0.1128	28.39	[30]
wood	101.51	0.0567	22.35	[31]
walnut shell-rice husk	126.72	0.0811	4.18	[32]

Table 1.
Textural properties of activated carbons from nanoparticles of agro wastes used as adsorbent.

Another prevailing quantitative characterization of nanoparticle from agrowaste is the Fourier transform infrared spectroscopy (FTIR). The principle involves passage of infrared (IR) radiations through the nanoparticle sample. Some of these IR radiations are absorbed by the material and the remaining transmitted. Absorption peaks at corresponding vibrations frequency between the atomic bonds are obtained in the resulting IR spectrum for interpretation. The peaks size in the spectrum gives information about the amount of material and chemical bonds present.

6. Adsorption of heavy metals from industrial wastewater using agro wastes nanoparticles

Adsorption process as industrial wastewater treatment had also proved to be superior to other conventional treatment techniques such as coagulation, flocculation, flotation, ion exchange, membrane filtration, electrochemical treatment, chemical precipitation and so on [33]. Adsorption involves binding of particles or molecules in a solution onto adsorbent surface (e.g. activated carbon from nanoparticles of agro wastes) and thus, it is usually a surface phenomenon. Heavy metals present in industrial wastewater can be removed from solution by adsorption process using activated carbon prepared from nanoparticles of agro wastes through calcination process. Advantages include adsorbent regeneration and reuse attribute, high adsorption efficiency, low cost of operation, heavy metals recovery possibility and low sludge formation [34, 35]. A temperature-controlled equipment with a stirring feature is usually used to study the batch adsorption process while concentration of heavy metals in solution is measured via atomic absorption spectrometer. Major factors that affect adsorption process of heavy metals from industrial wastewater using activated carbon (AC) prepared from nanoparticles of agro wastes include temperature, reaction time, AC dosage, stirring rate, solution pH and heavy metal concentration in solution [36]. The adsorption capacity of the AC from nanoparticles of agro wastes, q_e (mg/g) at equilibrium could be measured using Eq. (4).

$$q_e = (C_o - C_e)\frac{V}{W} \tag{4}$$

where C_o and C_e are initial and final concentrations of heavy metals in solution (mg/L), V is the volume of solution (L) and W is the weight of adsorbent (g).

Adsorption mechanisms of heavy metals over agro wastes nanoparticles from industrial wastewater include chemisorption, physisorption, complexation, ion exchange and pores diffusion through pores. Chemisorption is an irreversible process which involves chemical bonding of molecules in solution onto adsorbent surface while physisorption is a reversible process in which transfer or sharing of electrons does not occur.

7. Isotherm, kinetics and thermodynamics studies of heavy metals removal from industrial wastewater using activated carbon (AC) prepared from nanoparticles of agro wastes

Various isotherm, kinetics, mechanistic and thermodynamics models that can be used to study the adsorptive nature of activated carbon from nanoparticles of agro wastes on heavy metals in industrial wastewater are summarized in **Table 2**. The

Models	Equations	Applicability	Assumption
Isotherm Models			
Freundlich	$\log q_e = \log K_F + \frac{1}{n}\log C_e$	1. Heterogeneous surface energy systems. 2. Multilayer description adsorption with interaction between adsorbed molecules.	1. Adsorption of pollutant onto heterogeneous adsorbent
Langmuir	$\frac{C_e}{q_e} = \frac{1}{K_L q_{\max}} + \frac{C_e}{q_{\max}}$ $R_L = \frac{1}{1+K_L C_o}$	1. Monolayer coverage relationship formation of adsorbate molecules.	1. Existence of homogeneous active sites at adsorbent surface for monomolecular adsorbed layer formation. 2. Negligible interactions between adsorbed species.
Temkin	$q_e = b_T In A_T + b_T In C_e$	1. Covers adsorbate-adsorbent interaction.	1. Molecules heat adsorption decreases with adsorbent surface coverage increase. 2. Binding energies uniform distribution at the adsorbent surface.
Dubinin-Radushkevich	$In(q_e) = In(q_m) - B_D\varepsilon^2$ $\varepsilon = RTIn\left(1 + \frac{1}{C_e}\right)$ $E = \frac{1}{\sqrt{(-2B_D)}}$	1. Adsorbate-adsorbent equilibrium relation can be expressed independently of temperature. .	1. Adsorbent size is comparable to micropore size. 2. Characteristic curve Gaussian-type distribution
Harkin-Jura	$\frac{1}{q_e^2} = \frac{B_{HJ}}{A_{HJ}} - \left(\frac{1}{A_{HJ}}\right)\log C_e$		1. Multilayer adsorption on adsorbent surface with heterogeneous pore distribution.
Halsey	$Inq_e = \frac{1}{n_H} InK_H - \frac{1}{n_H} InC_e$	1. Multilayer adsorption	1. Heterogeneous adsorbent if equilibrium data are well fitted using Halsey isotherm.

Models	Equations	Applicability	Assumption
Redlich-Peterson	$In\left(K_{RP}\frac{C_e}{q_e}-1\right)=\beta_{RP}InC_e+Ina_{RP}$	1. Prediction of homogenous and heterogeneous adsorption systems.	
Sips	$In\left(\frac{q_e}{q_m-q_e}\right)=\frac{1}{n}In(C_e)+In(b_s)^{\frac{1}{n}}$	1. Merged Langmuir and Freundlich models. 2. Effectively reduced to Freundlich isotherm and disobey Henry's law at low adsorbate concentration. 3. Prediction of monolayer sorption capacity based on Langmuir isotherm at high concentrations of adsorbate.	
Toth	$In\left(\frac{q_e^{n_t}}{q_m^{n_t}-q_e^{n_t}}\right)=n_tInC_e+n_tInK_t$		
Adsorption Kinetics			
Pseudofirst-Order	$In(q_e-q_t)=Inq_e-k_1t$		
Pseudosecond-Order	$\frac{t}{q_t}=\frac{1}{k_2q_e^2}+\frac{t}{q_e}$		
Adsorption Mechanistic Modeling			
Intraparticle Diffusion	$q_t=K_{id}t^{0.5}+C$	1. Studies rate controlling step of biosorption process	
Adsorption Thermodynamics			
Thermodynamics	$\Delta G^o=-RTInK_L$ $\Delta G^o=\Delta H^o-T\Delta S^o$ $InK_L=\frac{\Delta S^o}{R}-\frac{\Delta H^o}{RT}$		

q_e ($mg\,g^{-1}$): Experimental adsorption capacity at equilibrium, K_F ($mg\,g^{-1}$) ($L\,mg^{-1}$)$^{1/?}$: Freundlich isotherm constant describing sorption capacity, C_e ($mg\,L^{-1}$): Adsorbate equilibrium concentration, n: a constant which gives information on heterogeneity grade, K_L ($L\,mg^{-1}$): Langmuir constant describing binding sites affinity and energy of adsorption, C_0 ($mg\,L^{-1}$): highest initial adsorbate concentration, R_L: dimensionless Langmuir equilibrium parameter, q_m ($mg\,g^{-1}$): adsorbent maximum monolayer adsorption capacity, R (8.314 $J\,mol^{-1}$): universal gas constant, T (°K): absolute temperature, b_T ($J\,mol^{-1}$): Temkin constant indicating heat of adsorption, A_T ($L\,mg^{-1}$): equilibrium binding constant related to maximum binding energy, B_D ($mol^2\,kJ^{-2}$): Dubinin-Radushkevich isotherm constant of adsorption energy, ?: Polanyi potential related to the equilibrium concentration, E ($kJ\,mol^{-1}$): mean free energy of adsorption, A_{HJ} and B_{HJ}: Harkin-Jurah adsorption constants; K_H and n_H: Halsey isotherm constants; K_{RP} (L/g): Redlich–Peterson isotherm constant, a_{RP} (L/mg): Redlich–Peterson isotherm constant, β: Redlich–Peterson exponent with values between 0 and 1, b_s: Sips isotherm constant related to energy of adsorption, K_t: Toth model adsorption isotherm constant, n_t: Toth model exponent, b_K: Khan model constant, a_K: Khan model exponent, q_t ($mg\,g^{-1}$): equilibrium adsorption uptake at time, t, k_1 (min^{-1}): pseudo-first-order rate constant of adsorption, k_2 ($g\,mg^{-1}\,min^{-1}$): pseudo-second-order rate constant of adsorption, K_{id} ($mg\,g^{-1}\,min^{-0.5}$): Intraparticle diffusion rate constant, C ($mg\,g^{-1}$): boundary layer diffusion effect, ΔG° ($kJ\,mol^{-1}$): change in Gibbs free energy, ΔH° ($kJ\,mol^{-1}$): change in enthalpy, ΔS° ($kJ\,mol^{-1}\,K^{-1}$): change in entropy, andK_{eq}: thermodynamic equilibrium constant.

Table 2.
Isotherm, kinetics, mechanistic and thermodynamics models for heavy metals [35].

adsorption isotherms explain the distribution of adsorbed molecules between industrial wastewater and activated carbon when equilibrium is attained during adsorption process. The isotherm models for fitting experimental data by linear regression method include two-parameters (Freundlich, Langmuir, Temkin, Dubinin-Radushkevich, Harkin-Jura and Halsey) and three-parameters (Redlich-Peterson, Sips and Toth) isotherm models. Adsorption kinetic studies predict the degree of heavy metals removal from industrial wastewater which governs adsorption reaction or residence time. The adsorption efficiency is a function of adsorbent adsorption potential. Adsorption mechanistic model gives information about the intraparticle diffusion of heavy metal molecules from bulk solution of industrial wastewater into pores of AC adsorbent. The correlation coefficient (R^2 value) then determines the isotherm, kinetics and mechanistic equations fitness for the adsorption process. Values of R^2 close to unity signify excellent fitness of models.

For thermodynamic analysis, the Gibbs free energy change (ΔG^o), enthalpy change (ΔH^o) and entropy change (ΔS^o) give information about inherent energetic changes during adsorption and show whether the reaction can happen spontaneously or not. Positive value of ΔG^o implies adsorption cannot happen spontaneously at examined temperatures. Also, values of ΔG^o between 0.0–20 kJ/mol and 80–400 kJ/mol respectively imply physical and chemical adsorption respectively. A positive value of enthalpy change (ΔH^o) suggests endothermic behavior of the

Error Function	Abbreviation	Model
Nonlinear chi-square test	χ^2	$\chi^2 = \sum_{i=1}^{n} \frac{(q_{e,\exp} - q_{e,calc})^2}{q_{e,\exp}}$
Sum of squares of the errors	*SSE*	$SSE = \sum_{i=1}^{n} (q_{e,\exp} - q_{e,calc})^2$
Average relative error	*ARE*	$ARE = \frac{100}{n} \sum_{i=1}^{n} \left\lvert \frac{q_{e,\exp} - q_{e,calc}}{q_{e,\exp}} \right\rvert$
Residual root mean square error	*RMSE*	$RMSE = \sqrt{\frac{1}{n-2} \sum_{i=1}^{n} (q_{e,\exp} - q_{e,calc})^2}$
Coefficient of determination	R^2	$R^2 = \frac{\sum (q_{e,\exp} - \bar{q}_{e,calc})^2}{\sum (q_{e,\exp} - \bar{q}_{e,calc})^2 + \sum (q_{e,\exp} - q_{e,cal})^2}$
Standard deviation of relative errors	S_{RE}	$S_{RE} = \sqrt{\frac{\sum_{i=1}^{n} [(q_{e,\exp} - q_{e,calc}) - ARE]^2}{n-1}}$
Marquardt's percent standard deviation	*MPSD*	$MPSD = 100 \sqrt{\frac{1}{n-p} \sum_{i=1}^{n} \left(\frac{q_{e,\exp} - q_{e,calc}}{q_{e,\exp}} \right)^2}$
Normalized standard deviation	*NSD*	$NSD = \Delta q(\%) = 100 \sqrt{\frac{1}{n-1} \sum_{i=1}^{n} \left(\frac{q_{e,\exp} - q_{e,calc}}{q_{e,\exp}} \right)^2}$
Hybrid functional error	*HYBRID*	$HYBRID = \frac{100}{(n-p)} \sum_{i=1}^{n} \frac{(q_{e,\exp} - q_{e,calc})}{q_{e,\exp}}$
Sum of absolute error	*EABS*	$EABS = \sum_{i=1}^{n} \lvert q_{e,\exp} - q_{e,calc} \rvert$

$q_{e,exp}$ *(mg g^{-1}): value obtained from the batch experiment,* $q_{e,calc}$ *(mg g^{-1}): calculated value from the isotherm for corresponding* $q_{e,exp}$, $\bar{q}_{e,calc}$ *(mg g^{-1}): mean of* $q_{e,exp}$, *n: number of experimental data points, and p: number of parameters in the respective model.*

Table 3.
List of error functions for isotherm and kinetics models fitness.

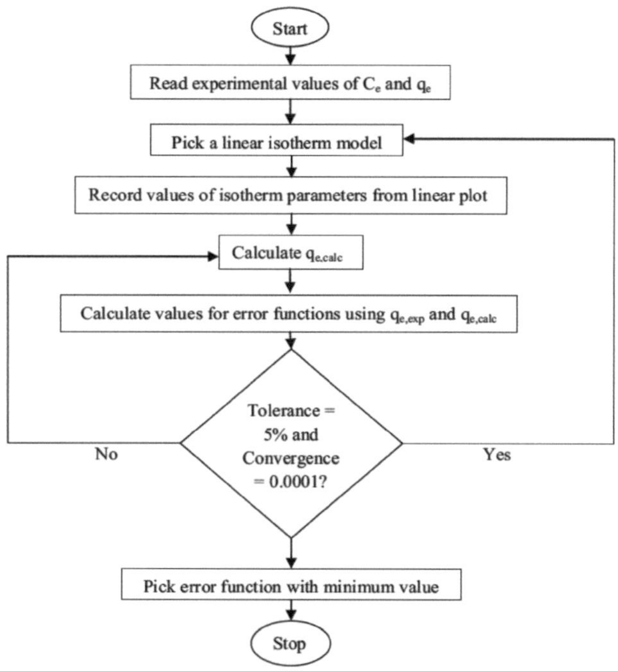

Figure 5.
Algorithm for linear isotherm models regression using error functions [38].

heavy metals adsorption process onto AC nanoparticles of agro wastes. Nevertheless, enthalpy change within 2.1–20.9, 20.9–80.0 and 80.0–418.4 kJ mol^{-1} implies physisorption, physisorption coupled with chemisorptions, and chemisorptions, respectively [37]. Positive value of entropy change (ΔS^o) indicates increase in the degree of freedom and disorderliness at solid–liquid interface during removal of heavy metals molecules on AC active sites.

The list of error functions that can be used to predict the degree of fitness of isotherm and kinetic models for heavy metals removal from industrial wastewater using AC from nanoparticles of agro wastes is presented in **Table 3**. The algorithms that can be used for both the linear and non-linear isotherm models regression are presented as **Figures 5** and **6** respectively [38].

8. Adsorption mechanism

The proposed mechanism of heavy metals (HMs) removal from industrial wastewater using activated carbon from nanoparticles of agro wastes (ACNAWs) involves binding electrostatic forces existence between positively charged HMs and existing negatively charged ions (-OH, -NH and COO^-) on ACNAWs. The ionization depends on the solution pH of the HMs which involves gaining or losing of a proton to solution. At low pH, ACNAWs become positively charged via gaining of a proton as written in Eqs. (5)–(7).

$$\text{ACNAW–OH} + H^+ \rightarrow \text{ACNAW–}OH_2{}^+ \quad (5)$$

$$\text{ACNAW–NH} + H^+ \rightarrow \text{ACNAW–}NH_2{}^+ \quad (6)$$

$$\text{ACNAW–COOH} + H^+ \rightarrow \text{ACNAW–}COOH_2{}^+ \quad (7)$$

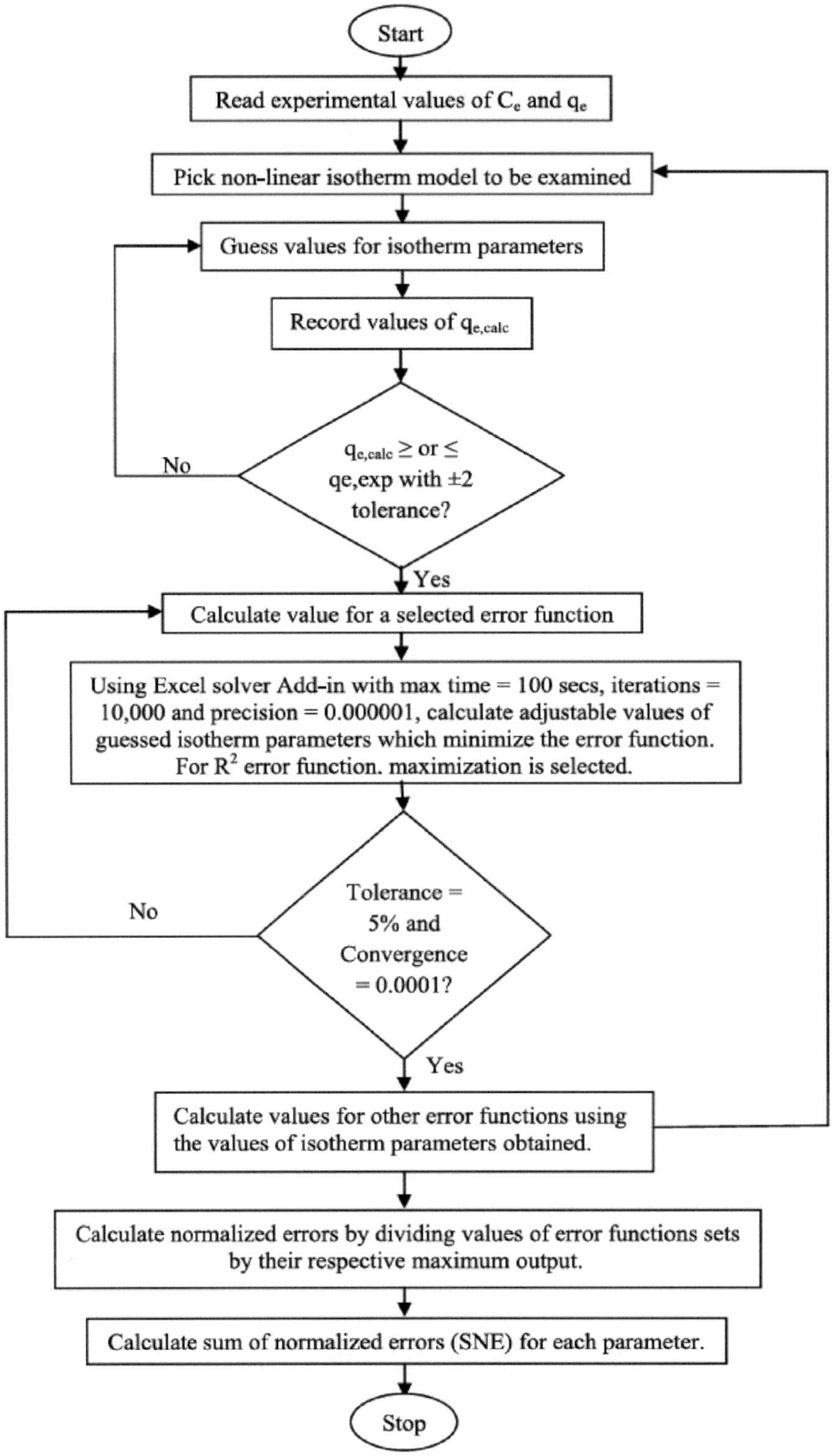

Figure 6.
Algorithm for non-linear isotherm models regression using error functions [38].

At very high pH, the surface of ACNAW becomes negatively charged via losing a proton as written in Eqs. (8)–(10).

$$\text{ACNAW–OH} + \text{OH}^- \rightarrow \text{ACNAW–O}^- + H_2O \tag{8}$$

$$\text{ACNAW–NH} + \text{OH}^- \rightarrow \text{ACNAW–N}^- + H_2O \tag{9}$$

$$\text{ACNAW–COOH} + \text{OH}^- \rightarrow \text{ACNAW–COO}^- + H_2O \tag{10}$$

The removal of HMs from industrial wastewater via adsorption onto ACNAW is then influenced by electrostatic attractive forces at high solution pH as written in Eqs. (11)–(13).

$$\text{ACNAW–O}^- + \text{HM} \rightarrow \text{ACNAW–O}^{-\cdots}\text{HM} \tag{11}$$

$$\text{ACNAW–N}^- + \text{HM} \rightarrow \text{ACNAW–N}^{-\cdots}\text{HM} \tag{12}$$

$$\text{ACNAW–COO}^- + \text{HM} \rightarrow \text{ACNAW–COO}^{-\cdots}\text{HM} \tag{13}$$

9. Conclusions

Heavy metals are hazardous substances which could be found in industrial wastewater effluents of industries. They cause havoc when absorbed into the body system. Also, wastes usually generated from agricultural practices pollute the environment and constitute nuisance. However, these agro wastes can be transformed into nanoparticles via mechanical technique and then to activated carbon via calcination process. These activated carbons can be used to remove heavy metals from industrial wastewater effluents via adsorption process. They have proved to be effective. Characterization techniques also revealed significant effects of calcination on structural, morphological and textural properties of activated carbon prepared from agro waste nanoparticles. Also, information obtained from isotherm, kinetics, mechanistic and thermodynamics studies give information about the adsorptive nature of the process. Conclusively, proposed adsorption mechanism is a key factor to know to how heavy metals are removed from industrial wastewater using activated carbons from nanoparticles of agro wastes.

Conflict of interest

No conflict of interest.

Author details

Lekan Taofeek Popoola[1*] and Alhaji Shehu Grema[2]

1 Chemical and Petroleum Engineering Department, Afe Babalola University, Ado-Ekiti, Ekiti State, Nigeria

2 Chemical Engineering Department, University of Maiduguri, Borno State, Nigeria

*Address all correspondence to: ltpopoola@abuad.edu.ng; popoolalekantaofeek@yahoo.com

References

[1] Ernhart CB. A critical review of low-level prenatal lead exposure in the human: 1. Effects on the fetus and newborn. Reprod. Toxicol. 1992; 6(1): 9-19.

[2] Khlifi R., Hamza-Chaffai A. Head and neck cancer due to heavy metal exposure via tobacco smoking and professional exposure: a review. Toxicol. Appl. Pharmacol. 2010; 248(2):71–88.

[3] Azimi A, Azari A, Rezakazemi M, Ansarpour M. Removal of heavy metals from industrial wastewaters: A review. Chemical and Biochemical Engineering Reviews. 2016; 4:37-59

[4] Gunatilake S. Methods of removing heavy metals from industrial wastewater. Journal of Multidisciplinary Engineering Science Studies. 2015; 1: 12-18

[5] Lasheen MR, El-Sherif IY, El-Wakee ST, Sabry DY, El-Shahat MF. Heavy metals removal from aqueous solution using magnetite Dowex 50WX4 resin nanocomposite. JMES. 2017; 8(2): 503-511.

[6] Shannon MA, Bohn PW, Elimelech M, Georgiadis JG, Marinas BJ, Mayes AM. Science and technology for water purification in the coming decades. Nature. 2008, 452, 301–310.

[7] Reddy DHK, Lee S, Seshaiah K. Removal of Cd(II) and Cu(II) from Aqueous Solution by Agro Biomass: Equilibrium, Kinetic and Thermodynamic Studies, Environ. Eng. Res., 2012; 17(3):125-132.

[8] FAO. Food wastage footprint impacts on natural resources: summary report. 2013. Food and Agriculture Organization of the United Nations, Rome.

[9] Kadirvelu K, Thamaraiselvi K, Namasivayam C. Removal of heavy metals from industrial wastewaters by adsorption onto activated carbon prepared from an agricultural solid waste. Bioresource Technology. 2001; 76(1): 63–65.

[10] Renu MA, Singh K, Upadhyaya S, Dohare R. Removal of heavy metals from wastewater using modified agricultural adsorbents. Mater. Today Proc. 2017; 4(10):534–538

[11] Huang J, Cao Y, Liu Z, Deng Z, Tang F. Efficient removal of heavy metal ions from water system by titanate nanoflowers. Chem. Eng. J. 2012; 180:75–80.

[12] Bailey SE, Olin TJ, Bricka M, Adrian DD. A review of potentially low-cost sorbents for heavy metals. Water Res. 1999; 33:2469–2479.

[13] Adelere IA, Lateef A. A novel approach to the green synthesis of metallic nanoparticles: the use of agro-wastes, enzymes and pigments, Nanotechnol Rev. 2016; 5(6):567–587

[14] Laurent S, Forge D, Port M, Roch A, Robic C, Vander Elst L, Muller RN. Magnetic iron oxide nanoparticles: synthesis, stabilization, vectorization, physicochemical characterizations, and biological applications. Chem. Rev. 2010; 110.

[15] Tiwari JN, Tiwari RN, Kim KS. Zero-dimensional, one-dimensional, two-dimensional and three-dimensional nanostructured materials for advanced electrochemical energy devices. Prog. Mater Sci. 2012; 57: 724-803.

[16] Shin,W.-K, Cho J, Kannan AG, Lee Y.-S, Kim D.-W. Cross-linked composite gel polymer electrolyte using mesoporous methacrylate-functionalized SiO_2 nanoparticles for lithium-ion polymer batteries. Sci. Rep. 2016; 6, 26332.

[17] Bhatnagar A, Sillanpaa M. Utilization of agro-industrial and municipal waste materials as potential adsorbents for water treatment: a review. Chem. Eng. J. 2010; 157:277–296.

[18] Pokhrel M, Wahid K, Mao Y. Systematic studies on RE_2-Hf_2O_7:5%Eu^{3+} (RE = Y, La, Pr, Gd, Er, and Lu) nanoparticles: effects of the A-site RE^{3+} cation and calcination on structure and photoluminescence. J. Phys. Chem. C 2016; 120:14828-14839.

[19] Dash S, Kamruddin M, Ajikumar P, Tyagi A, Raj B. Nanocrystalline and metastable phase formation in vacuum thermal decomposition of calcium carbonate. Thermochimica Acta, 2000; 363(1):129–135.

[20] Chandrasekhar S, Pramada PN, Majeed J. Efect of calcination temperature and heating rate on the optical properties and reactivity of rice husk ash. J Mater Sci. 2006; 41(23): 7926–7933

[21] Ullah H, Khan I, Yamani ZH, Qurashi A. Sonochemicaldriven ultrafast facile synthesis of SnO_2 nanoparticles: growth mechanism structural electrical and hydrogen gas sensing properties. Ultrason. Sonochem. 2017; 34:484-490.

[22] Jenkins R, Snyder R. Introduction to X-ray Powder Diffractometry, 2012; Vol. 267, Wiley, Interscience.

[23] Hammond C, Hammond C. The Basics of Crystallography and Diffraction, 2009; 3rd edition, Oxford, United States, Oxford University Press Inc., New York.

[24] Iqbal N, Khan I, Yamani ZH, Qurashi A. Sonochemical assisted solvothermal synthesis of gallium oxynitride nanosheets and their solar-driven photoelectrochemical water-splitting applications. Sci. Rep. 2016; 6, 32319.

[25] Popoola LT, Aderibigbe TA, Yusuff AS. Synthesis of composite snail shell-rice husk adsorbent for brilliant green dye uptake from aqueous solution: Optimization and Characterization. Environmental Quality Management (Published online). 2019; DOI: https://doi.org/10.1002/tqem.21602.

[26] Lykhach Y, Kozlov SM, Skála T, Tovt A, Stetsovych V, Tsud N, Dvořák F, Johanek V, Neitzel A, Myslivecek J, Fabris S, Matolın V, Neyman KM, Libuda J. Counting electrons on supported nanoparticles. Nat. Mater. 2015; http://dx.doi.org/10.1038/nmat4500.

[27] Kestens V, Roebben G, Herrmann J, Jämting A, Coleman V, Minelli C, Clifford C, De Temmerman P.-J, Mast J, Junjie L, Babick F, Colfen H, Emons H. Challenges in the size analysis of a silica nanoparticle mixture as candidate certified reference material. J. Nanopart. Res. 2016; 18, 171.

[28] Brunauer S, Emmett P, Teller E. Adsorption of gases in multimolecular layers. Journal of the American Chemical Society, 1938; 60(2):309–319.

[29] Zainol MM, Asmadi M, Amin NAS. Preparation and Characterization of Impregnated Magnetic Particles on Oil Palm Frond Activated Carbon for Metal Ions Removal. Sains Malaysiana, 2017; 46(5):773-782.

[30] Idan IJ, Luqman CA, Thomas SYC, Siti NABMJ. Equilibrium, kinetics and thermodynamic adsorption studies of acid dyes on adsorbent developed from kenaf core fiber. Adsorption Science and Technology, 2018; 36(1–2):694-712.

[31] Sahu JN, Acharya J, Meikap BC. Response surface modeling and optimization of chromium(VI) removal from aqueous solution using tamarind wood activated carbon in batch process. Journal of Hazardous Materials, 2009; 172(2-3):818-825.

[32] Popoola LT. Nano-magnetic walnut shell-rice husk for Cd(II) sorption: design and optimization using artificial intelligence and design expert. Heliyon. 2019; 5:e02381.

[33] Feng L, Cao M, Ma X, Zhu Y, Hu C. Superparamagnetic high-surface-area Fe_3O_4 nanoparticles as adsorbents for arsenic removal. Journal of Hazardous Materials. 2012; 217, 439–446.

[34] Burakov AE, Galunin EV, Burakova IV, Kucherova AE, Agarwal S, Tkachev AG, Gupta VK. Adsorption of heavy metals on conventional and nanostructured materials for wastewater treatment purposes: a review. Ecotoxicol. Environ. Saf. 2018; 148:702–712.

[35] Popoola LT, Aderibigbe TA, Yusuff AS, Munir MM. Brilliant green dye adsorption onto composite snail shell –rice husk: Adsorption isotherm, kinetic, mechanistic and thermodynamic analysis. Environmental Quality management. 2018. (published online)DOI: https://doi.org/10.1002/tqem.21597.

[36] Popoola LT Tetracycline and sulfamethoxazole adsorption onto nanomagnetic walnut shell-rice husk: isotherm, kinetic, mechanistic and thermodynamic studies. International Journal of Environmental Analytical Chemistry. 2019a; Article in Press. DOI: 10.1080/03067319.2019.1646739

[37] Yalcin S. The mechanism of heavy metal biosorption on green marine macroalga Enteromorpha linza. CLEAN-Soil, Air,Water. 2014; 42(3): 251-259.

[38] Popoola LT. Characterization and adsorptive behaviour of snail shell-rice husk (SS-RH) calcined particles (CPs) towards cationic dye. Heliyon. 2019b; 5: e01153.doi:10.1016/j.heliyon.2019.e01153.

Chapter 6

Porous ZnO Nanostructures Synthesized by Microwave Hydrothermal Method for Energy Harvesting Applications

Sofia Henriques Ferreira, Ana Rovisco, Andreia dos Santos, Hugo Águas, Rui Igreja, Pedro Barquinha, Elvira Fortunato and Rodrigo Martins

Abstract

The ever-growing global market for smart wearable technologies and Internet of Things (IoT) has increased the demand for sustainable and multifunctional nanomaterials synthesized by low-cost and energy-efficient processing technologies. Zinc oxide (ZnO) is a key material for this purpose due to the variety of facile methods that exist to produced ZnO nanostructures with tailored sizes, morphologies, and optical and electrical properties. In particular, ZnO nanostructures with a porous structure are advantageous over other morphologies for many applications because of their high specific surface area. In this chapter, a literature review on the latest progress regarding the synthesis and applications of ZnO with a porous morphology will be provided, with special focus on the synthesis by microwave hydrothermal method of these nanomaterials and their potential for application in energy harvesting devices. Nanogenerators of a composite made by polydimethylsiloxane (PDMS) and porous ZnO nanostructures were explored and optimized, with an output voltage of (4.5 ± 0.3) V being achieved for the best conditions. The daily life applicability of these devices was demonstrated by lighting up a commercial LED, by manually stimulating the nanogenerator directly connected to the LED.

Keywords: zinc oxide, microwave synthesis, porous nanostructures, energy harvesting devices

1. Introduction

Zinc oxide (ZnO) is an inorganic semiconductor material that has been applied in a wide range of applications over the last centuries [1]. The attraction to ZnO can be attributed to its remarkable optical and electronic characteristics. With a direct and wide bandgap of 3.37 eV and a large exciton binding energy of 60 meV at room temperature [2], ZnO has the potential to be applied in advanced electronic and optoelectronic devices with promising results, such as UV sensors [3, 4], transparent electrodes [5, 6], gas sensors [7], thin film transistors [8, 9], and solar cells [10–12].

Moreover, ZnO is a low-cost and biocompatible material with high photostability, high chemical and thermal stability, low toxicity, and a broad range of UV radiation absorption [13]. These properties allow ZnO to be applied in a wide range of applications besides electronic devices, such as skin ointments and sunscreens, rubber tires, paints, bioimaging, drug delivery, biosensors, antibacterial textiles, and photocatalysis for the degradation of pollutants in wastewaters [1, 14–19].

Due to its piezoelectric properties, ZnO nanostructures have also been widely explored for energy harvesting applications, being an important sustainable energy source [20]. The demand for wearable devices led to a high development of new energy sources. Nanogenerators have demonstrated the capability to power small electronic devices, appearing as a good alternative to batteries [21]. The most common nanogenerators are based on piezoelectric and/or triboelectric effects. In the piezoelectric nanogenerators, mechanical energy is converted into electrical energy through piezoelectric polarization resultant from strain [1]. The triboelectric effect results from the surface charges' generation subsequent from the friction between two different materials (with opposite triboelectric polarities) [22].

Materials with piezoelectric properties have the capability to convert mechanical energy into electrical energy [1]. Within the different piezoelectric materials, lead zirconate titanate (PZT) is the material that presented so far the highest piezoelectric coefficient (d_{33} = 593 pC N^{-1}), still this material has a high toxicity [23, 24]. While presenting a much lower d_{33} value($\approx$ 10 pC N^{-1}) [25–27], ZnO is a very good alternative, since it is not only sustainable and eco-friendly, as it can also be easily fabricated, while still presenting a good performance [28, 29].

Nanogenerators of different types of ZnO nanostructures (i.e., nanorods, nanoparticles, nanoflowers) have been reported [30–34]. For example, Saravanakumar et al. reported a nanogenerator fabricated using vertically grown ZnO nanowires with surrounding PDMS, with output values of 6 V/4 nA/0.39 nW cm^{-2} under finger bending [35]. Rahman et al. used ZnO nanoparticles dispersed into a PDMS film, achieving output values of 20 V/20 μA/20 μW, with finger tapping [36]. As another example, ZnO nanoflowers were mixed with multiwalled carbon nanotubes and PDMS, with an output of 75 V/3.2 μA/260 mW cm^{-2} being obtained. In this case, the devices were tested in the soles of human shoes with the force being applied by a person walking [37].

1.1 Synthesis and applications of porous ZnO nanostructures

Despite all the established applications of ZnO, the research involving this semiconductor has not yet diminished, mostly due to the continuing development of new synthesis technologies and applications. For instance, ZnO nanomaterials can be easily synthesized into tailored sizes and morphologies at low temperatures (< 200 °C) by a variety of methods, including chemical bath deposition [38], electrodeposition [39], chemical vapor deposition [40], electrospinning [41], laser assisted flow deposition [42], and solvothermal [16] or hydrothermal [43, 44] synthesis, either by conventional or microwave-assisted heating [4, 45].

Porous oxide semiconductor nanomaterials, particularly two-dimensional (2D) materials with nanoscale thickness, are promising candidates due to their usually large specific surface areas that can improve their performance in several applications [46–51]. These nanomaterials can inclusively assemble into three-dimensional (3D) hierarchically structures with controlled morphology and dimensions which can lead to novel properties and applications [52]. The self-assemble technique is a facile method to produce 3D hierarchical structures where low-dimension building units aggregate spontaneously into high-dimensional architectures. This technique offers many advantages over other methods as it can be performed at low temperatures using low-cost materials while having high yield for scale production [52].

An indirect way to produce porous 3D ZnO structures has been recently developed by thermal decomposition of layered zinc hydroxide (LZH) precursors [52]. LZHs are usually composed of positively charged zinc hydroxyl layers intercalated by anions that balance the overall charge and water molecules [53]. The anions in LZH generally include CO_3^{2-}, SO_4^{2-}, NO_3^-, Cl^-, CH_3COO^- [50, 53–58]. These LZH precursors are fabricated with the desired morphology and then converted into porous ZnO nanomaterials by a calcination process at high temperatures [59]. During calcination, the precursors release gaseous molecules and, consequently, the original structure contracts and pores are formed throughout the structures [52].

LZHs are typically obtained via solution techniques, mainly hydrothermal methods where the materials' synthesis occurs in a basic medium that results from the addition of certain reagents, such as hexamethylenetetramine [60], ammonia [58, 61], and urea [52, 62–65]. Although the basic structure is similar in all LZHs, the sites occupied by the anions and water molecules are different and, as a result, the final morphology, crystal structure, interlayer distances, and thermal decomposition temperature differ depending on the anion type [66]. In particular, the LZH carbonate (LZHC) is composed of zinc hydroxide layers combined with carbonate ions and water molecules. During the synthesis of this material, a well-crystallized phase is typically obtained with an invariable distance between the LZH [53]. The resulting morphology of LZHC usually consists of 2D structures stacked in a hierarchical 3D arrangement. However, the synthesis of uniform LZHC 3D morphologies through a simple and fast hydrothermal method has not yet been fully explored.

For this purpose, hydrothermal synthesis assisted by microwave irradiation offers many advantages over conventional heating. In a synthesis assisted by conventional heating, the heat transfer occurs through a combination of conductive and convective mechanisms that result in a low heating rate and, consequently, long synthesis time [67]. Conventional heating method is also dependent on the thermal conduction of the material of which the reaction vessel walls are made. Moreover, the temperature maximum occurs on the vessel wall surface, as shown in **Figure 1**. All these factors can lead to a non-uniform heating of the reaction medium and, subsequently, originate a heterogeneity in the obtained products [1]. On the other hand, hydrothermal synthesis assisted by microwave irradiation allows for rapid and uniform heating since the heat transfer occurs directly from the microwaves to the molecules of the reaction's materials, as illustrated in **Figure 1**. This results in high reaction rates and a homogeneous and volumetric heating [68, 69].

The porous morphology of ZnO nanostructures obtained by calcination of LZHC significantly increases the materials' specific surface area [70] and, therefore,

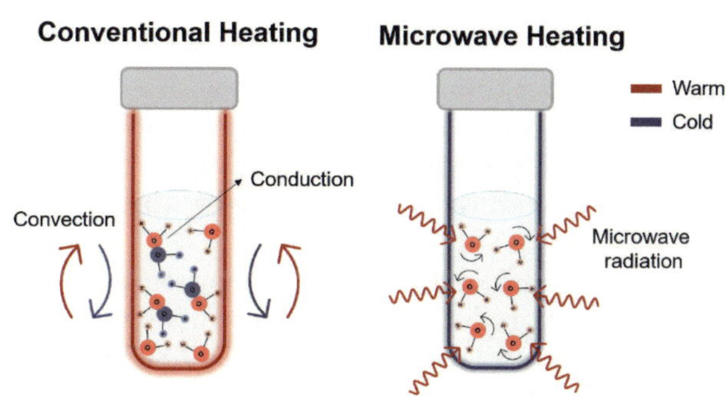

Figure 1.
Schematic of conventional heating versus microwave heating processes.

these ZnO nanomaterials have been used in applications that benefit from this characteristic, such as photocatalysis [51, 52, 70], gas sensors [50, 54, 71–73], surface enhanced Raman scattering (SERS) substrates [74], dye-synthesized solar cells [44, 75, 76], and battery electrodes [65].

This work aims to demonstrate the potential of high surface area porous ZnO nanostructures for energy harvesting devices, showing original and novel results regarding the characterization of nanogenerators based on these structures. For that, 3D hierarchically structures composed of LZHC nanoplates were successfully synthesized through a facile, low-cost, and low temperature hydrothermal process assisted by microwave irradiation. Porous ZnO nanostructures were obtained by calcination of the LZHC at 700 °C for 2 h in air while maintaining the LZHC hierarchical 3D structure. Porous ZnO nanostructures were then embedded in PDMS and deposited by spin-coating technique on flexible substrates. Energy harvesting based on a micro-structured composite of porous ZnO nanostructures embedded in PDMS was investigated. The combination of using the porous ZnO nanostructures, which have piezoelectric properties, and triboelectricity resultant from the micro-structuring leads to a performance improvement of the nanogenerators [37, 77]. To the best of our knowledge, porous ZnO nanostructures were for the first time used to fabricate a micro-structured PDMS/ZnO composite for energy harvesting devices.

2. Materials and methods

2.1 Synthesis and characterization of porous ZnO nanostructures

Porous ZnO nanostructures were synthesized by hydrothermal method assisted by microwave irradiation. Zinc nitrate hexahydrate ($Zn(NO_3)_2{\cdot}6H_2O$, Sigma-Aldrich 98%) and urea (CH_4N_2O, Sigma-Aldrich 99.0–100.5%) were used without further purification. In a typical synthesis, 0.05 M of zinc nitrate was first dissolved in de-ionized water, and after its total dissolution, urea was added to the aqueous solution. The molar ratio of zinc to urea was kept at 1:5. Then, 25 mL of the obtained solution was transferred to a 35 mL Pyrex vessel which was placed in a CEM Discovery SP microwave. The synthesis was carried out at 140 °C for 15 min under a power of 100 W.

After the synthesis, the resulting white precipitates were washed with de-ionized water followed by isopropanol and centrifuged at 4500 rpm for 5 min. This washing process was repeated three times. The powders were dried in air at room temperature for 48 h and then calcinated in air in a Nabertherm muffle furnace at 700 °C for 2 h with a heating rate of 250 °C h^{-1}.

The crystallinity of the produced nanostructures was analyzed by X-ray diffraction (XRD) using a PANalytical's X'Pert PRO MRD X-ray diffractometer, with a monochromatic Cu Kα radiation source with wavelength 1.540598 Å. XRD measurements were carried out from 10 to 90° (2θ), with a scanning step size of 0.016°. The morphology of the LZHC precursor and porous ZnO nanostructures was evaluated by scanning electron microscopy (SEM) using a Carl Zeiss AURIGA CrossBeam FIB-SEM workstation equipped with an Oxford X-ray Energy Dispersive Spectrometer.

Differential scanning calorimetric (DSC) and thermogravimetry (TG) measurements of the synthesized product without any temperature treatment were carried out with a simultaneous thermal analyzer NETZSCH STA 449 F3 Jupiter. Approximately 20 mg of the synthesized powder was loaded into an open platinum-rhodium crucible and heated in air from room temperature to 850 °C with a heating rate of 10 °C min^{-1}.

Diffuse reflectance measurements of the porous ZnO nanostructures were performed at room temperature using a PerkinElmer lambda 950 UV/VIS/NIR spectrophotometer with a diffuse reflectance module with a 150 mm diameter integrating sphere, internally coated with Spectralon. The calibration of the system was achieved by using a standard Spectralon reflector sample as reference. The reflectance spectra were obtained from 350 to 800 nm.

2.2 Fabrication and characterization of energy harvesting devices

The devices were fabricated as described in references [27, 78] and the fabrication process is illustrated in **Figure 2**. Briefly, composites of porous ZnO nanostructures embedded in PDMS were produced with concentrations of 20, 25, and 30 wt%. Firstly, the nanostructures were mixed with the PDMS elastomer (from Dow Corning) and a volume of ethyl acetate (from Fluka-Honeywell) enough to ensure a homogeneous mixture of elastomer and nanostructures. The mixture was stirred until the evaporation of the solvent, and then a curing agent (Sylgard 184, from Dow Corning) was added in a weight ratio to the elastomer of 1:10 while stirring to obtain a homogeneous mixture. Two types of devices were produced, unstructured and micro-structured nanogenerators. The former was fabricated by spin-coating the mixture at 250 rpm for 90 s, with an acceleration of 100 rpm·s^{-1}, on commercial substrates of polyethylene terephthalate (PET) with a layer of indium tin oxide (ITO) deposited on top (PET/ITO, from Kintec Company), whereas the latter was obtained by depositing the mixture in a similar way on acrylic molds (5 mm thick, from Dagol). The acrylic molds used were produced as described in reference [79].

The composites were then cured at 60 °C for 1 h. After the curing process, PET/ITO electrodes were placed on top of the composite films, as shown in **Figure 2**. The electrical characterization of the produced nanogenerators was performed by applying a mechanical stimulus in a contact area of 0.3 cm^2 with a pushing force of 2.3 N at different frequencies (0.5, 1, 1.5, and 2 pushes per second) with a home-made machine with a linear motor.

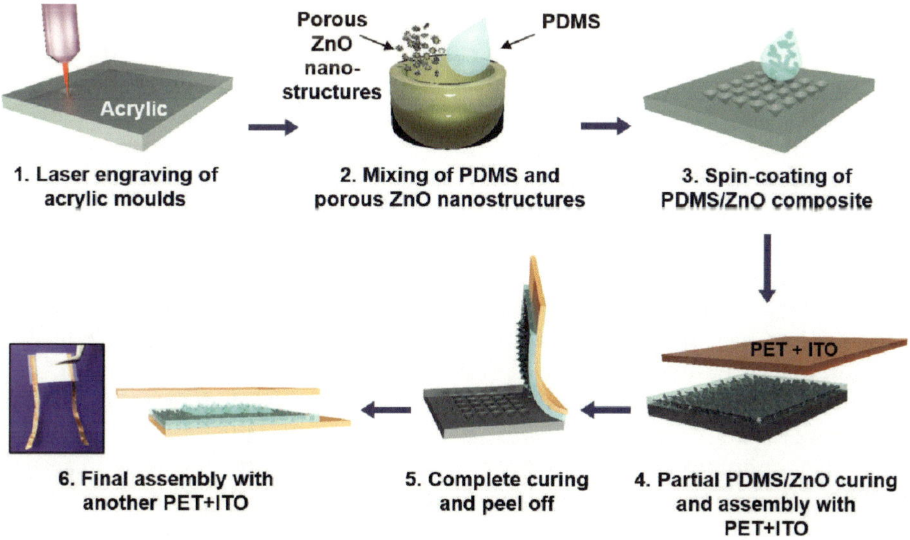

Figure 2.
Fabrication schematic of a micro-structured nanogenerator based on a PDMS/ZnO composite film. Adapted from dos Santos et al. [78].

3. Porous ZnO nanostructures as a piezoelectric material for nanogenerators

3.1 Synthesis of porous ZnO nanostructures

3.1.1 Characterization of the LZHC precursor

Figure 3(a) presents the X-ray diffractogram obtained from the final product of the hydrothermal synthesis prior to the calcination process. All the peaks from the diffractogram can be indexed to zinc hydroxide carbonate hydrate ($Zn_4(CO_3)(OH)_6{\cdot}H_2O$) (ICDD 11–0287). The morphology of the precursor was observed in SEM and it is shown in **Figure 3(b)**. The SEM image reveals that the LZHC precursor obtained after only 15 min of microwave hydrothermal synthesis consists of many flower-like structures, with a few micrometers of diameter, composed of densely packed LZHC nanoplates with a few nanometers of thickness.

Differential scanning calorimetric (DSC) measurements were carried out in air from room temperature to 800 °C to analyze the conversion process of LZHC into ZnO. The DSC curve in **Figure 3(c)** shows two endothermic peaks at 64 °C and 266 °C. The peak at 64 °C corresponds to the removal of water that is weakly adsorbed to the LZHC nanostructures [80], resulting in a weight loss of 38.12%. The second peak at 266 °C results in a weight loss of 14.87% and it is associated with the release of water and carbon dioxide from the thermal decomposition of LZHC precursor [50, 59].

3.1.2 Characterization of the porous ZnO nanostructures

After the calcination process, the LZHC precursor was successfully converted into porous ZnO nanostructures, which can be inferred from the X-ray diffractograms of the samples obtained at 700 °C, depicted in **Figure 4(a)**. All the peaks in the diffractogram correspond to the hexagonal wurtzite ZnO structure (ICDD 36–1451). No characteristic peaks from any other impurities were detected, indicating that the LZHC precursor was completely converted into ZnO. SEM images of the calcinated product are presented in **Figure 4(b)** with different magnifications. The low magnification image shows that the morphology of the final ZnO product did not suffer significant changes when compared with the LZHC precursor, since ZnO nanoplates are still assembled into flower-like structures. However, when observing the high magnification SEM images, it is possible to see that the ZnO nanoplates present a porous structure with serrate edges and a wide pore size distribution.

The synthesis of LZHC precursor by urea-assisted hydrothermal method, followed by the calcination process to originate porous ZnO nanostructures, has been explained before in the literature [44, 65, 73, 80, 81]. **Figure 5** shows a simple schematic of the synthesis and transformation process of LZHC precursor into porous ZnO nanostructures. During the hydrothermal synthesis, urea is hydrolyzed leading to the formation of hydroxide (OH^-) and carbonate (CO_3^{2-}) ions. Zn^{2+} ions from the added zinc salt react with both OH^- and CO_3^{2-} ions forming the LZHC precursor ($Zn_4(CO_3)(OH)_6{\cdot}H_2O$). It has been reported that the surface of LZHC plates is hydrophobic whereas the lateral sides are hydrophilic, resulting in a vertical growth of this material and consequent plate-like morphology [76, 82, 83]. The agglomeration of these nanoplates into stable flower-like microstructures occurs to favor the minimization of surface energy by reducing exposed surface areas [83]. Under calcination at high temperature, LZHC is decomposed into ZnO by releasing

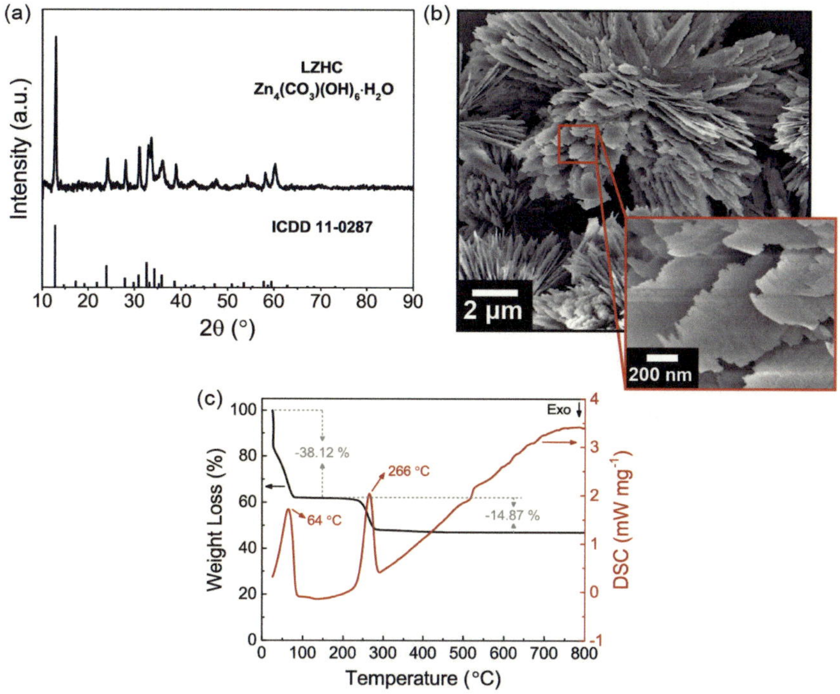

Figure 3.
(a) X-ray diffractogram, (b) SEM images and (c) TG/DSC curves of the LZHC precursor.

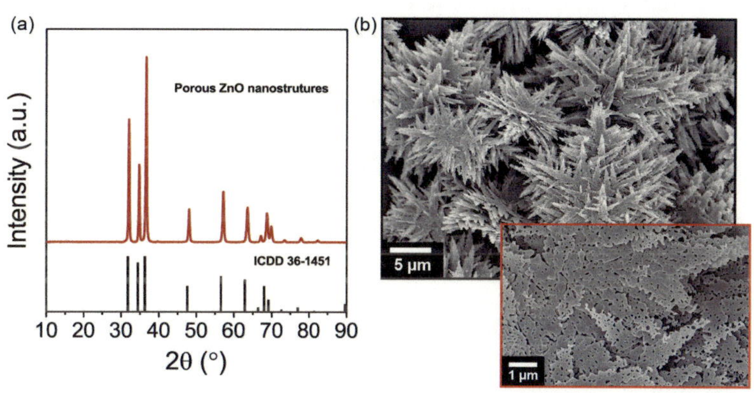

Figure 4.
(a) XRD diffractogram and (b) SEM images of porous ZnO nanostructures synthesized by hydrothermal method assisted by microwave irradiation followed by calcination at 700 °C for 2 h in air.

H_2O and CO_2 in the form of gas, which leads to a contraction of the original structure which originates pores throughout the nanoplates and a consequent porous ZnO structure, as illustrated in **Figure 5** [84].

The UV–Vis diffuse reflectance of the produced ZnO samples is presented in **Figure 6**. The optical band gap E_g was calculated by applying the Kubelka-Munk (K-M) method to the reflectance (R) data [85]. The K-M method is based on the following equation:

$$F(R) = \frac{(1-R)^2}{2R} \tag{1}$$

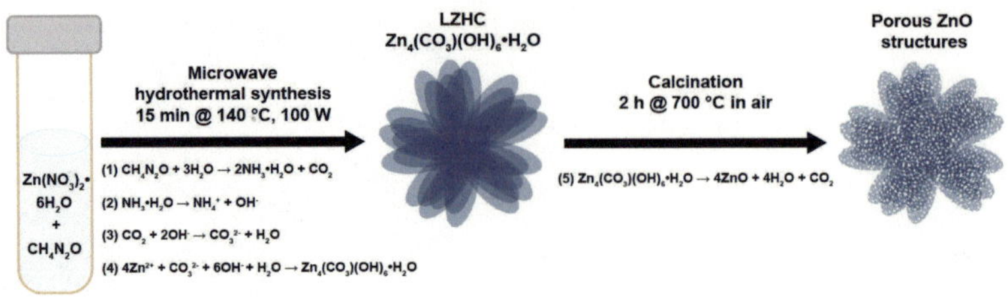

Figure 5.
Schematic of the hydrothermal synthesis assisted by microwave irradiation and calcination process of porous ZnO structures.

The K-M function ($F(R)$) is proportional to the absorption coefficient (α). Therefore, by considering the Tauc relation, the following expressions can be obtained [86]:

$$F(R) \propto \alpha \propto = \frac{\left(h\upsilon - E_g\right)^{1/n}}{h\upsilon} \tag{2}$$

$$\left(F(R)h\upsilon\right)^n = A\left(h\upsilon - E_g\right) \tag{3}$$

where A is a constant and n is equal to 2 for semiconductors with direct allowed transitions [87]. As shown by the inset graph in **Figure 6**, the value of E_g can be determined by extrapolating the linear part of the function curve with the energy axis. The estimated bandgap energy is 3.26 eV for ZnO nanostructures obtained at 700 °C, which is consistent with the values reported in the literature [52, 88].

3.2 Fabrication of energy harvesting devices

3.2.1 Characterization of the PDMS/ZnO composite films

Composites of porous ZnO nanostructures embedded in PDMS (PDMS/ZnO films) were fabricated. The composites were produced with a micro-structuring and in an unstructured form. SEM images of a micro-structured porous PDMS/ZnO film are presented in **Figure 7(a)**. The array of aligned cones observed has an average height of 380 μm, an average diameter of 300 μm, and a gap around 100 μm. **Figure 7(b)** combines the XRD diffractogram of the porous ZnO nanostructures, the PDMS/ZnO composite film, and the pure PDMS film. As expected, even if presenting a much lower intensity, the hexagonal wurtzite ZnO structure (ICDD 36–1451) can be identified in the PDMS/ZnO composite, whereas the PDMS film presents an amorphous structure.

3.2.2 Performance of the PDMS/ZnO nanogenerators

To optimize the nanogenerator output, its performance was evaluated by varying the concentration of the porous ZnO nanostructures in the PDMS film. This study was performed with unstructured composites. Three concentrations were considered to produce the devices: 20, 25, and 30 wt%. **Figure 8(a)** presents the peak-to-peak output voltage of the nanogenerators. The electrical characterization of the nanogenerators was performed by applying a mechanical stimulus with a pushing force of 2.3 N at frequency of 2 pushes per second with a home-made

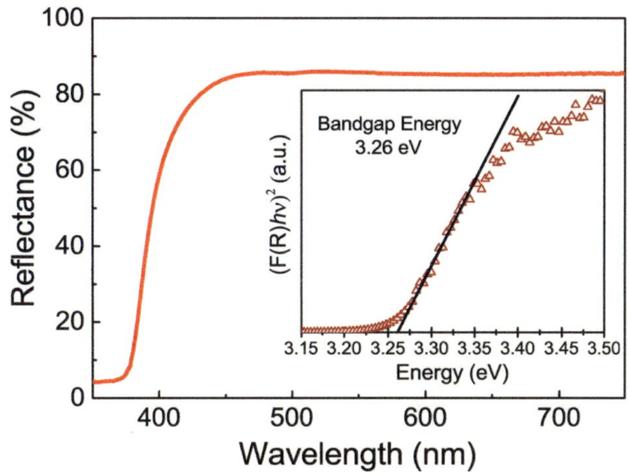

Figure 6.
Reflectance spectra of the porous ZnO nanostructures with an inset graphic showing the obtained bandgap energy via the K-M function.

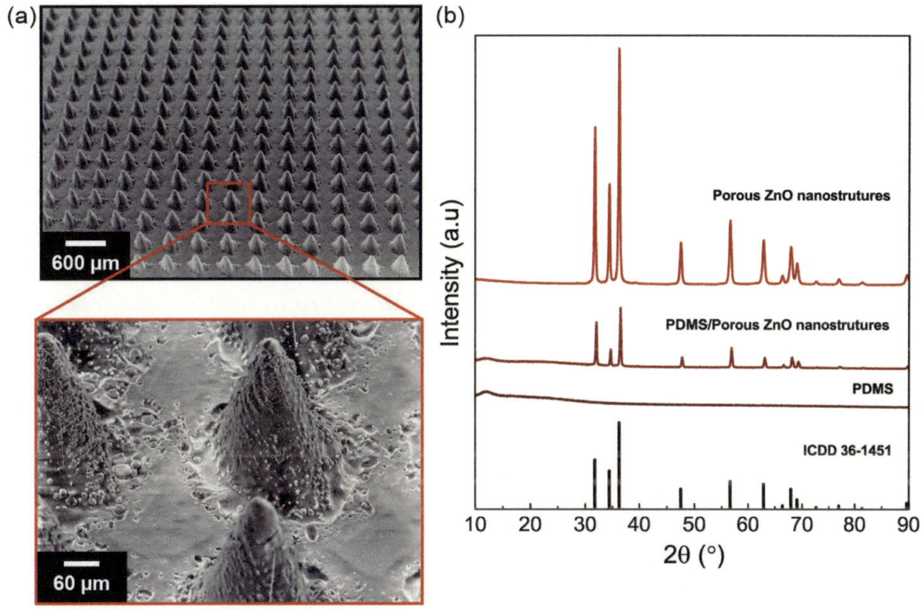

Figure 7.
(a) SEM images of a micro-structured PDMS/ZnO composite film, with the insets displaying closer views of the micro-cones. (b) XRD diffractogram of porous ZnO nanostructures, PDMS/ZnO composite film, and PDMS film.

bending machine. The obtained results reveal an increase of the output voltage from 20 to 25 wt%, and then a decrease for 30 wt%. These results are in agreement to what was previously observed using the same approach for ZnO nanorods, where the optimal concentration for the nanogenerators output was also 25 wt% [78]. As such, to further characterize the nanogenerators, the concentration considered was 25 wt%.

In previous studies from our group [27, 78], an enhanced response was achieved by micro-structuring the composite, as shown in **Figure 7**, and, therefore, the same approach was adopted in this study. **Figure 8(b)** presents the output voltage

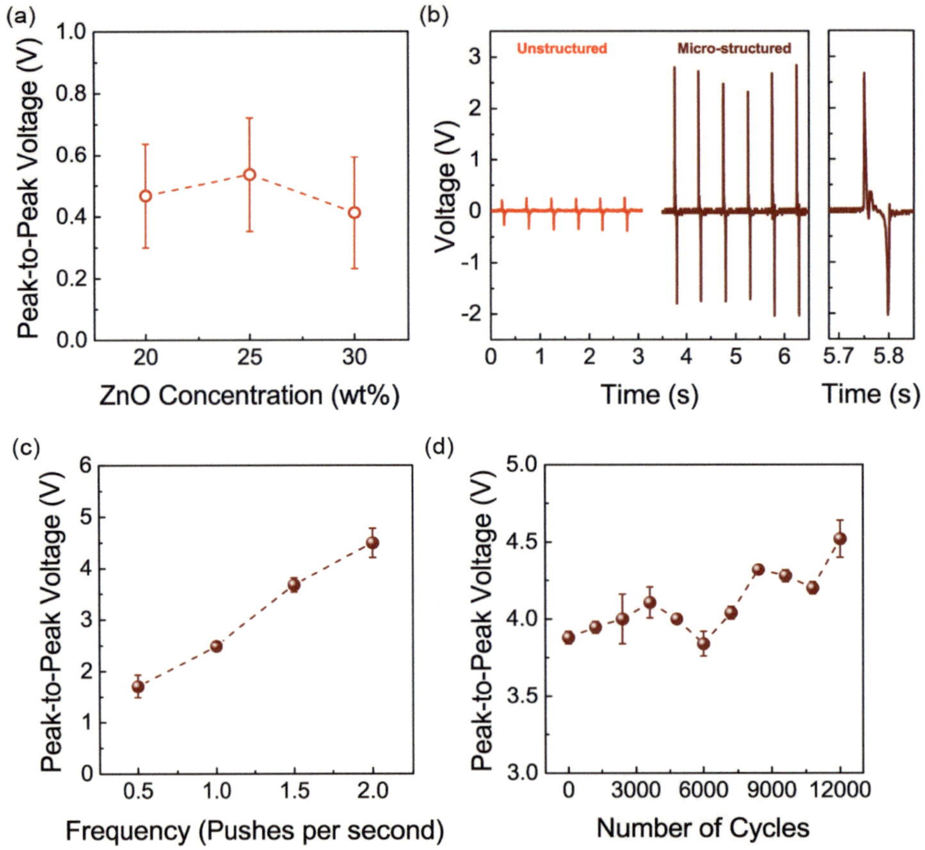

Figure 8.
(a) Peak-to-peak output voltage for PDMS/ZnO composites with different concentrations of porous ZnO nanostructures. Note that each point was determined using the average output of 2–6 equal devices. (b) Output voltage for an unstructured and a micro-structured nanogenerator with a porous ZnO nanostructures concentration of 25 wt%. (c) Peak-to-peak voltage for different frequencies applying a pushing force of 2.3 N. (d) Output voltage from the optimized nanogenerator for 12,000 cycles.

for this nanogenerator in comparison with the unstructured one. A peak-to-peak output voltage of (4.5 ± 0.3) V was obtained for the micro-structured nanogenerator against only (0.5 ± 0.2) V for the unstructured one. The micro-structuring can not only improve the force delivery into the nanostructures, leading to an increase of the piezoelectric effect, but it can also induce an extra triboelectric effect, as a consequence of the air gaps between the PDMS/ZnO composite micro-structures and the ITO electrode. These two effects originate an enhanced response of the micro-structured nanogenerator.

Considering the micro-structured nanogenerator with the best performance (25 wt%), the influence of varying the frequency of the stimulus was investigated while maintaining the applied force at 2.3 N. **Figure 8(c)** shows the peak-to-peak output voltage of the nanogenerator as function of the frequency, where the output voltage increases with increasing frequency. This trend has been observed by other groups, and it can be explained by the eventual accumulation of residual charges due to an inefficient neutralization of the induced charges provoked by a faster stimulation [89].

To study the potential of the nanogenerator in a daily life application, its stability along 12,000 cycles was also investigated. For this study, the stimulus was applied with a pushing force of 2.3 N while maintaining the frequency at 2 pushes per second. **Figure 8(d)** shows the output voltage along the pushing cycles, and no deterioration of its performance is observed. Instead, it is possible to detect a slight increase of the output voltage to (7.2 ± 0.1) V along the pushing cycles, which can also be related to charges accumulation.

3.2.3 Proof-of-concept of the PDMS/ZnO nanogenerator

To understand the applicability of the micro-structured PDMS/ZnO nanogenerator, it is important to study its performance when connected to external load resistances with different values (1 to 30 MΩ). This study was performed with a fixed pushing of 2.3 N at 2 pushes per second. **Figure 9(a)** presents the peak-to-peak output voltage and current, while **Figure 9(b)** shows the resultant instantaneous power density. An increase of the power density with increasing load resistance is observed until 10 MΩ, reaching a maximum value of 2.7 μW cm^{-2}, after that a slight decrease is observed. Comparing to the recent results on PDMS/ZnO nanorods nanogenerators [78], the maximum power obtained here is just slightly lower, presenting the same order of magnitude. Its lower output is expected due the absence of a preferential direction for piezoelectric response (c-axes) in these nanostructures. Nevertheless, the synthesis of these porous ZnO nanostructures allows for a faster and low-cost fabrication of nanogenerators, since it is a rapid, simple, and high yield approach to obtain ZnO nanostructures.

Additionally, the nanogenerator output is very satisfactory, proven to be enough to light up a blue LED (2.8–4 V, 20 mA), by directly connecting the nanogenerator to the LED and manually stimulating the energy harvester, as shown in **Figures 9(c)** and **(d)** and Video 1 available from (can be viewed at) https://youtu.be/JCT60ozKCX8. These results prove not only the applicability of these

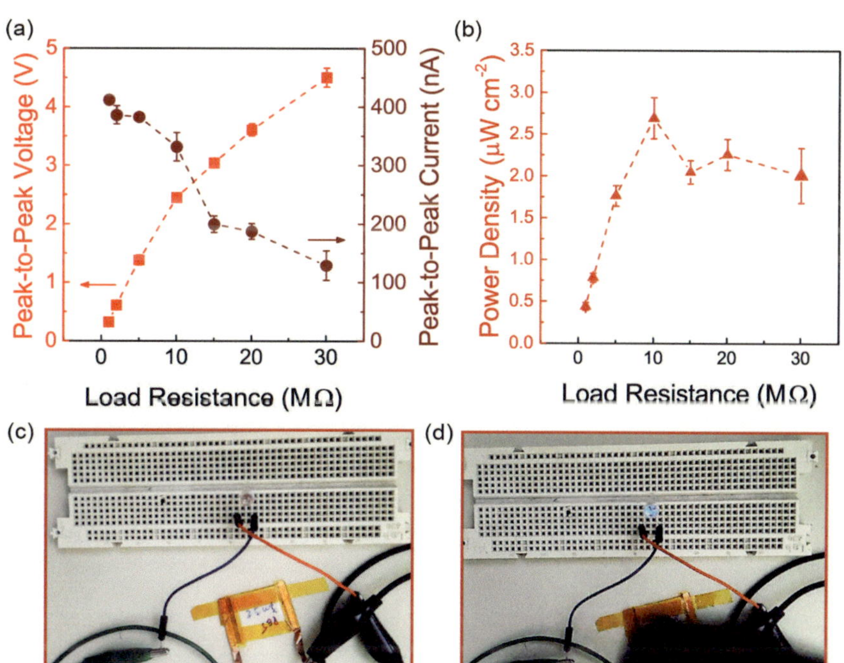

Figure 9.
Application of the micro-structured PDMS/ZnO nanogenerator, stimulated with a pushing force of 2.3 N at a frequency of 2 pushes per second varying the load resistance. Peak-to-peak (a) voltage and current outputs, and (b) correspondent power density for several load resistances. Note that each peak-to-peak value is an average of 5 measurements. (c and d) Nanogenerator directly lighting up a blue LED by applying manual force.

nanogenerators in simple daily life applications but also demonstrate their potential to power wearable sensors or multifunctional platforms where these porous ZnO nanostructures are employed in more than one application.

4. Conclusions

In summary, porous ZnO nanostructures were successfully synthesized via a facile and fast hydrothermal method assisted by microwave irradiation and calcinated at 700 °C for 2 h in air. The effect of calcination temperature on the morphological, structural, and optical properties of the porous ZnO nanostructures was investigated. Nanogenerators based on a micro-structured composite of PDMS with embedded porous ZnO structures were successfully produced, reaching an output voltage of (4.5 ± 0.3) V. The devices proved to be very robust and stable by presenting no deterioration of their performance after 12,000 pushing cycles. An external load of 10 MΩ optimized the nanogenerators performance, reaching a power density of 2.7 μW cm^{-2}. The capability of these nanogenerators to lighting up commercial LEDs, through direct connection and with a manual stimulus, was shown, demonstrating their potential for daily life applications.

Acknowledgements

This work is funded by National Funds through FCT - Portuguese Foundation for Science and Technology, Reference UIDB/50025/2020-2023 and FCT/MCTES. This work also received funding from the European Community's H2020 programme under grant agreement No. 787410 (ERC-2018-AdG DIGISMART), No. 716510 (ERC-2016-StG TREND) and No. 952169 (SYNERGY, H2020-WIDESPREAD-2020-5, CSA). S. H. F. acknowledges the Portuguese Foundation for Science and Technology for the AdvaMTech PhD program scholarship PD/BD/114086/2015 and IDS-FunMat-INNO-2 project FPA2016/EIT/EIT RawMaterials Grant Agreement 17184.

Conflict of interest

The authors declare no conflict of interest.

Notes/thanks/other declarations

The authors would like to thank Ana Pimentel for the TG-DSC measurement and Daniela Nunes for the SEM images.

Author details

Sofia Henriques Ferreira*, Ana Rovisco*, Andreia dos Santos, Hugo Águas, Rui Igreja, Pedro Barquinha, Elvira Fortunato and Rodrigo Martins*
CENIMAT/i3N, Department of Materials Science, NOVA School of Science and Technology (FCT-NOVA) and CEMOP/UNINOVA, NOVA University Lisbon, Caparica, Portugal

*Address all correspondence to: sdl.ferreira@campus.fct.unl.pt; a.rovisco@campus.fct.unl.pt and rm@uninova.pt

References

[1] Wojnarowicz J, Chudoba T, Lojkowski W. A Review of Microwave Synthesis of Zinc Oxide Nanomaterials : Reactants , Process Parameters and Morphologies. Nanomaterials. 2020;10:1086.

[2] Wang ZL. Zinc oxide nano structures: Growth, properties and applications. J Phys Condens Matter. 2004;16:R829–R858.

[3] Samouco A, Marques AC, Pimentel A, Martins R, Fortunato E. Laser-induced electrodes towards low-cost flexible UV ZnO sensors. Flex Print Electron. 2018;3:No. 044002.

[4] Ferreira SH, Deuermeier J, Sequeira S, Nunes D, Gonçalves AMF, Martins R, et al. Industrial waste residue converted into value-added ZnO for optoelectronic applications. ACS Appl Electron Mater. 2020;2(7):1960-1969.

[5] Lyubchyk A, Vicente A, Soule B, Alves PU, Mateus T, Mendes MJ, et al. Mapping the Electrical Properties of ZnO-Based Transparent Conductive Oxides Grown at Room Temperature and Improved by Controlled Postdeposition Annealing. Adv Electron Mater. 2016;2:1500287.

[6] Fortunato E, Fortunato E, Nunes P, Nunes P, Marques a, Marques a, et al. Transparent, conductive ZnO:Al thin film deposited on polymer substrates by RF magnetron sputtering. Science (80-). 2002;152:247-51.

[7] Zhu L, Zeng W. Room-temperature gas sensing of ZnO-based gas sensor: A review. Sensors Actuators, A Phys. 2017;267:242-261.

[8] Sun Y, Maemoto T, Sasa S. Fully transparent ZnO thin-film transistors using conducting AZO films fabricated at room temperature. 2014;2-3.

[9] Fortunato EMC, Barquinha PMC, Pimentel ACMBG, Gonçalves AMF, Marques AJS, Pereira LMN, et al. Fully transparent ZnO thin-film transistor produced at room temperature. Adv Mater. 2005;17:590-594.

[10] Muchuweni E, Sathiaraj TS, Nyakotyo H. Hydrothermal synthesis of ZnO nanowires on rf sputtered Ga and Al co-doped ZnO thin films for solar cell application. J Alloys Compd. 2017;721:45-54.

[11] Cao J, Wu B, Chen R, Wu Y, Hui Y, Mao BW, et al. Efficient, Hysteresis-Free, and Stable Perovskite Solar Cells with ZnO as Electron-Transport Layer: Effect of Surface Passivation. Adv Mater. 2018;30(11):1-9.

[12] Nayeri FD, Soleimani EA, Salehi F. Synthesis and characterization of ZnO nanowires grown on different seed layers: The application for dye-sensitized solar cells. Renew Energy. 2013;60:246-255.

[13] Özgür Ü, Alivov YI, Liu C, Teke a., Reshchikov M a., Doğan S, et al. A comprehensive review of ZnO materials and devices. J Appl Phys. 2005;98(4):1-103.

[14] Ong CB, Ng LY, Mohammad AW. A review of ZnO nanoparticles as solar photocatalysts: Synthesis, mechanisms and applications. Renew Sustain Energy Rev. 2018;81(March 2017):536-551.

[15] Samadi M, Zirak M, Naseri A, Khorashadizade E, Moshfegh AZ. Recent progress on doped ZnO nanostructures for visible-light photocatalysis. Thin Solid Films. 2016;605:2-19.

[16] Pimentel A, Rodrigues J, Duarte P, Nunes D, Costa FM, Monteiro T, et al. Effect of solvents on ZnO nanostructures synthesized

by solvothermal method assisted by microwave radiation: a photocatalytic study. J Mater Sci. 2015;50:5777-5787.

[17] Theerthagiri J, Salla S, Senthil RA, Nithyadharseni P, Madankumar A, Arunachalam P, et al. A review on ZnO nanostructured materials : energy, environmental and biological applications. Nanotechnology. 2019;30:392001.

[18] Rong P, Ren S, Yu Q. Fabrications and Applications of ZnO Nanomaterials in Flexible Functional Devices-A Review. Crit Rev Anal Chem. 2019;49(4):336-349.

[19] Sanguino P, Monteiro T, Bhattacharyya SR, Dias CJ, Igreja R, Franco R. ZnO nanorods as immobilization layers for interdigitated capacitive immunosensors. Sensors Actuators, B Chem. 2014;204:211-217.

[20] Askari H, Khajepour A, Khamesee MB, Saadatnia Z, Wang ZL. Piezoelectric and triboelectric nanogenerators: Trends and impacts. Nano Today. 2018 Oct;22(1):10-13.

[21] Wang ZL. Nanogenerators for Self-powered Devices and Systems. 2011.

[22] Indira SS, Vaithilingam CA, Oruganti KSP, Mohd F, Rahman S. Nanogenerators as a sustainable power source: state of art, applications, and challenges. Nanomaterials. 2019;9(5):773.

[23] Safari A, Akdogan EK. Piezoelectric and Acoustic Materials for Transducer Applications. 2008.

[24] Ibn-Mohammed T, Reaney IM, Koh SCL, Acquaye A, Sinclair DC, Randall CA, et al. Life cycle assessment and environmental profile evaluation of lead-free piezoelectrics in comparison with lead zirconate titanate. J Eur Ceram Soc. 2018;38:4922-4938.

[25] Fan HJ, Lee W, Hauschild R, Alexe M, Rhun G Le, Scholz R, et al. Template-assisted large-scale ordered arrays of ZnO pillars for optical and piezoelectric applications. Small. 2006;2(4):561-568.

[26] Broitman E, Soomro MY, Lu J, Willander M, Hultman L. Nanoscale piezoelectric response of ZnO nanowires measured using a nanoindentation technique. Phys Chem Chem Phys. 2013 Jul;15(26):11113-11118.

[27] Rovisco A, dos Santos A, Cramer T, Martins J, Branquinho R, Águas H, et al. Piezoelectricity Enhancement of Nanogenerators Based on PDMS and ZnSnO 3 Nanowires through Microstructuration. ACS Appl Mater Interfaces. 2020 Apr;12(16):18421-18430.

[28] Shetti NP, Bukkitgar SD, Reddy KR, Reddy CV, Aminabhavi TM. ZnO-based nanostructured electrodes for electrochemical sensors and biosensors in biomedical applications. Biosens Bioelectron. 2019;141(March):111417.

[29] Yin B, Qiu Y, Zhang H, Lei J, Chang Y, Ji J, et al. Piezoelectric effect of 3-D ZnO nanotetrapods. RSC Adv. 2015;5(15):11469-11474.

[30] Wu JM, Xu C, Zhang Y, Yang Y, Zhou Y, Wang ZL. Flexible and transparent nanogenerators based on a composite of lead-free ZnSnO3 triangular-belts. Adv Mater. 2012;24(45):6094-6099.

[31] Guo R, Guo Y, Duan H, Li H, Liu H. Synthesis of Orthorhombic Perovskite-Type ZnSnO3 Single-Crystal Nanoplates and Their Application in Energy Harvesting. ACS Appl Mater Interfaces. 2017 Mar;9(9):8271-8279.

[32] Alam MM, Ghosh SK, Sultana A, Mandal D. Lead-free ZnSnO3/MWCNTs-based self-poled flexible hybrid nanogenerator for

piezoelectric power generation. Nanotechnology. 2015;26(16):165403.

[33] Choi KH, Siddiqui GU, Yang B, Mustafa M. Synthesis of ZnSnO3 nanocubes and thin film fabrication of (ZnSnO3/PMMA) composite through electrospray deposition. J Mater Sci Mater Electron. 2015;26:5690-5696.

[34] Paria S, Karan SK, Bera R, Das AK, Maitra A, Khatua BB. A Facile Approach To Develop a Highly Stretchable PVC/ ZnSnO 3 Piezoelectric Nanogenerator with High Output Power Generation for Powering Portable Electronic Devices. Ind Eng Chem Res. 2016 Oct;55(40):10671-10680.

[35] Saravanakumar B, Mohan R, Thiyagarajan K, Kim SJ. Fabrication of a ZnO nanogenerator for eco-friendly biomechanical energy harvesting. RSC Adv. 2013;3(37):16646-16656.

[36] Rahman W, Garain S, Sultana A, Ranjan Middya T, Mandal D. Self-Powered Piezoelectric Nanogenerator Based on Wurtzite ZnO Nanoparticles for Energy Harvesting Application. Materials Today: Proceedings. 2018. p. 9826-9830.

[37] Kim DH, Dudem B, Yu JS. High-Performance Flexible Piezoelectric-Assisted Triboelectric Hybrid Nanogenerator via Polydimethylsiloxane-Encapsulated Nanoflower-like ZnO Composite Films for Scavenging Energy from Daily Human Activities. ACS Sustain Chem Eng. 2018;6(7):8525-8535.

[38] Manthina V, Agrios AG. Single-pot ZnO nanostructure synthesis by chemical bath deposition and their applications. Nano-Structures & Nano-Objects. 2016 Jul 1;7:1-11.

[39] Ait hssi A, Amaterz E, Labchir N, Atourki L, Bouderbala IY, Elfanaoui A, et al. Electrodeposited ZnO Nanorods as Efficient Photoanodes for the Degradation of Rhodamine B. Phys Status Solidi. 2020;217(17):2000349.

[40] Lupan O, Emelchenko GA, Ursaki V V., Chai G, Redkin AN, Gruzintsev AN, et al. Synthesis and characterization of ZnO nanowires for nanosensor applications. Mater Res Bull. 2010;45(8):1026-1032.

[41] Lin D, Wu H, Pan W. Photoswitches and memories assembled by electrospinning aluminum-doped zinc oxide single nanowires. Adv Mater. 2007;19(22):3968-3972.

[42] Rodrigues J, Fernandes AJS, Monteiro T, Costa FM. A review on the laser-assisted flow deposition method: growth of ZnO micro and nanostructures. CrystEngComm. 2019;21(7):1071-1090.

[43] Mahpeykar SM, Koohsorkhi J, Ghafoori-fard H. Ultra-fast microwave-assisted hydrothermal synthesis of long vertically aligned ZnO nanowires for dye-sensitized solar cell application. Nanotechnology. 2012;23(16):165602.

[44] Qiu Y, Chen W, Yang S. Facile hydrothermal preparation of hierarchically assembled, porous single-crystalline ZnO nanoplates and their application in dye-sensitized solar cells. J Mater Chem. 2010;20(5):1001-1006.

[45] Pimentel A, Nunes D, Duarte P, Rodrigues J, Costa FM, Monteiro T, et al. Synthesis of long ZnO nanorods under microwave irradiation or conventional heating. J Phys Chem C. 2014;118:14629-14639.

[46] Jang JS, Lee SE, Choi SJ, Koo WT, Kim DH, Shin H, et al. Heterogeneous, Porous 2D Oxide Sheets via Rapid Galvanic Replacement: Toward Superior HCHO Sensing Application. Adv Funct Mater. 2019;29(42):1-10.

[47] Huang A, He Y, Zhou Y, Zhou Y, Yang Y, Zhang J, et al. A review of recent

applications of porous metals and metal oxide in energy storage, sensing and catalysis. J Mater Sci. 2019;54(2):949-973.

[48] Sk MM, Yue CY, Ghosh K, Jena RK. Review on advances in porous nanostructured nickel oxides and their composite electrodes for high-performance supercapacitors. J Power Sources. 2016;308:121-140.

[49] Butburee T, Bai Y, Wang H, Chen H, Wang Z, Liu G, et al. 2D Porous TiO2 Single-Crystalline Nanostructure Demonstrating High Photo-Electrochemical Water Splitting Performance. Adv Mater. 2018;30(21):1-8.

[50] Xie X, Wang X, Tian J, Song X, Wei N, Cui H. Growth of porous ZnO single crystal hierarchical architectures with ultrahigh sensing performances to ethanol and acetone gases. Ceram Int. 2017;43(1):1121-1128.

[51] Jin Z, Zhang YX, Meng FL, Jia Y, Luo T, Yu XY, et al. Facile synthesis of porous single crystalline ZnO nanoplates and their application in photocatalytic reduction of Cr(VI) in the presence of phenol. J Hazard Mater. 2014;276:400-407.

[52] Liu S, Li C, Yu J, Xiang Q. Improved visible-light photocatalytic activity of porous carbon self-doped ZnO nanosheet-assembled flowers. CrystEngComm. 2011;13(7):2533-2541.

[53] Song B, Wang Y, Cui X, Kou Z, Si L, Tian W, et al. A Series of Unique Architecture Building of Layered Zinc Hydroxides: Self-Assembling Stepwise Growth of Layered Zinc Hydroxide Carbonate and Conversion into Three-Dimensional ZnO. Cryst Growth Des. 2016;16(2):887-894.

[54] Zhang L, Zhao J, Lu H, Li L, Zheng J, Li H, et al. Facile synthesis and ultrahigh ethanol response of hierarchically porous ZnO nanosheets. Sensors Actuators, B Chem. 2012;161(1):209-215.

[55] Moezzi A, Cortie M, McDonagh A. Transformation of zinc hydroxide chloride monohydrate to crystalline zinc oxide. Dalt Trans. 2016;45(17):7385-7390.

[56] Moezzi A, Cortie MB, McDonagh AM. Zinc hydroxide sulphate and its transformation to crystalline zinc oxide. Dalt Trans. 2013;42(40):14432-14437.

[57] Hosono E, Fujihara S, Kimura T, Imai H. Growth of layered basic zinc acetate in methanolic solutions and its pyrolytic transformation into porous zinc oxide films. J Colloid Interface Sci. 2004;272(2):391-398.

[58] Song RQ, Xu AW, Deng B, Li Q, Chen GY. From layered basic zinc acetate nanobelts to hierarchical zinc oxide nanostructures and porous zinc oxide nanobelts. Adv Funct Mater. 2007;17(2):296-306.

[59] Li B, Wang Y. Hierarchically assembled porous ZnO microstructures and applications in a gas sensor. Superlattices Microstruct. 2011;49(4):433-440.

[60] Mao J, Li JJ, Ling T, Liu H, Yang J, Du XW. Facile synthesis of zinc hydroxide carbonate flowers on zinc oxide nanorods with attractive luminescent and optochemical performance. Nanotechnology. 2011;22(24):245607.

[61] Chen M, Wang Z, Han D, Gu F, Guo G. Porous ZnO polygonal nanoflakes: Synthesis, use in high-sensitivity NO 2 gas sensor, and proposed mechanism of gas sensing. J Phys Chem C. 2011;115(26):12763-12773.

[62] Bitenc M, Marinšek M, Crnjak Orel Z. Preparation and characterization

of zinc hydroxide carbonate and porous zinc oxide particles. J Eur Ceram Soc. 2008;28(15):2915-2921.

[63] Gu F, You D, Wang Z, Han D, Guo G. Improvement of gas-sensing property by defect engineering in microwave-assisted synthesized 3D ZnO nanostructures. Sensors Actuators, B Chem. 2014;204(3):342-350.

[64] Zhang Y, Liu C, Gong F, Jiu B, Li F. Large scale synthesis of hexagonal simonkolleit nanosheets for ZnO gas sensors with enhanced performances. Mater Lett. 2017;186(August 2016):7-11.

[65] Teng Y, Mo M, Li Y, Xue J, Zhao H. Amorphous carbon-coated ZnO porous nanosheets: Facile fabrication and application in lithium- and sodium-ion batteries. J Alloys Compd. 2018;744:712-720.

[66] Shinagawa T, Watanabe M, Mori T, Tani JI, Chigane M, Izaki M. Oriented Transformation from Layered Zinc Hydroxides to Nanoporous ZnO: A Comparative Study of Different Anion Types. Inorg Chem. 2018;57(21):13137-13149.

[67] Pimentel A, Nunes D, Pereira S, Martins R, Fortunato E. Photocatalytic Activity of TiO2 Nanostructured Arrays Prepared by Microwave-Assisted Solvothermal Method. Cao W, editor. Semiconductor Photocatalysis - Materials, Mechanisms and Applications. 2016. p. 81-103.

[68] Hayes BL. Microwave Synthesis: Chemistry at the Speed of Light. 2002.

[69] Mirzaei A, Neri G. Microwave-assisted synthesis of metal oxide nanostructures for gas sensing application: A review. Sensors Actuators, B Chem. 2016;237:749-775.

[70] Liu D, Lv Y, Zhang M, Liu Y, Zhu Y, Zong R, et al. Defect-related photoluminescence and photocatalytic properties of porous ZnO nanosheets. J Mater Chem A. 2014;2(37):15377-15388.

[71] Zhou J, Gong F, Wang H, Xiao Y, Li F, Mai W. 3D mace-like hierarchical ZnO nanoarchitecture constructed with microrod bundles and porous single-crystalline nanosheets for acetone sensors with enhanced performances. Mater Sci Eng B Solid-State Mater Adv Technol. 2017;225(May):68-74.

[72] Lin Z, Guo F, Wang C, Wang X, Wang K, Qu Y. Preparation and sensing properties of hierarchical 3D assembled porous ZnO from zinc hydroxide carbonate. RSC Adv. 2014;4(10):5122-5129.

[73] Chang J, Ahmad MZ, Wlodarski W, Waclawik ER. Self-assembled 3D ZnO porous structures with exposed reactive {0001} facets and their enhanced gas sensitivity. Sensors. 2013;13(7):8445-8460.

[74] Yang H, Ni S-Q, Jiang X, Jiang W, Zhan J. In situ fabrication of single-crystalline porous ZnO nanoplates on zinc foil to support silver nanoparticles as a stable SERS substrate. CrystEngComm. 2012;14(18):6023.

[75] Lai YH, Lin CY, Chen HW, Chen JG, Kung CW, Vittal R, et al. Fabrication of a ZnO film with a mosaic structure for a high efficient dye-sensitized solar cell. J Mater Chem. 2010;20(42):9379-9385.

[76] Hosono E, Fujihara S, Honma I, Zhou H. The fabrication of an upright-standing zinc oxide nanosheet for use in dye-sensitized solar cells. Adv Mater. 2005;17(17):2091-2094.

[77] Zhang XS, Han M Di, Wang RX, Zhu FY, Li ZH, Wang W, et al. Frequency-multiplication high-output triboelectric nanogenerator for sustainably powering biomedical microsystems. Nano Lett. 2013;13(3):1168-1172.

[78] dos Santos A, Sabino F, Rovisco A, Barquinha P, Águas H, Fortunato E, et al. Optimization of ZnO Nanorods Concentration in a Micro-Structured Polymeric Composite for Nanogenerators. Chemosensors. 2021 Jan;9(2):27.

[79] dos Santos A, Pinela N, Alves P, Santos R, Fortunato E, Martins R, et al. Piezoresistive E-Skin Sensors Produced with Laser Engraved Molds. Adv Electron Mater. 2018;4(9):1800182-1800192.

[80] Jing Z, Zhan J. Fabrication and Gas-Sensing Properties of Porous ZnO Nanoplates. Adv Mater. 2008;(20):4547-4551.

[81] Liu X, Zhang J, Wang L, Yang T, Guo X, Wu S, et al. 3D hierarchically porous ZnO structures and their functionalization by Au nanoparticles for gas sensors. J Mater Chem. 2011;21(2):349-356.

[82] Yang H, Ni SQ, Jiang X, Jiang W, Zhan J. In situ fabrication of single-crystalline porous ZnO nanoplates on zinc foil to support silver nanoparticles as a stable SERS substrate. CrystEngComm. 2012;14(18):6023-6028.

[83] Xingfu Z, Zhaolin H, Yiqun F, Su C, Weiping D, Nanping X. Microspheric organization of multilayered ZnO nanosheets with hierarchically porous structures. J Phys Chem C. 2008;112(31):11722-11728.

[84] Wang X, Cai W, Lin Y, Wang G, Liang C. Mass production of micro/nanostructured porous ZnO plates and their strong structurally enhanced and selective adsorption performance for environmental remediation. J Mater Chem. 2010;20(39):8582-8590.

[85] Mishra V, Warshi MK, Sati A, Kumar A, Mishra V, Sagdeo A, et al. Diffuse reflectance spectroscopy: An effective tool to probe the defect states in wide band gap semiconducting materials. Mater Sci Semicond Process. 2018;86:151-156.

[86] Akir S, Barras A, Coffinier Y, Bououdina M, Boukherroub R, Omrani AD. Eco-friendly synthesis of ZnO nanoparticles with different morphologies and their visible light photocatalytic performance for the degradation of Rhodamine B. Ceram Int. 2016;42(8):10259-10265.

[87] Viezbicke BD, Patel S, Davis BE, Birnie DP. Evaluation of the Tauc method for optical absorption edge determination: ZnO thin films as a model system. Phys Status Solidi Basic Res. 2015;252(8):1700-1710.

[88] Huang N, Shu J, Wang Z, Chen M, Ren C, Zhang W. One-step pyrolytic synthesis of ZnO nanorods with enhanced photocatalytic activity and high photostability under visible light and UV light irradiation. J Alloys Compd. 2015;648:919-929.

[89] Ko YH, Nagaraju G, Lee SH, Yu JS. PDMS-based Triboelectric and Transparent Nanogenerators with ZnO Nanorod Arrays. ACS Appl Mater Interfaces. 2014 May;6(9):6631-6637.

Chapter 7

Nanoporous Metallic Films

Swastic and Jegatha Nambi Krishnan

Abstract

Nanoporous metallic films are known to have high surface to volume ratio due to the presence of pores. The presence of pores and ligaments make them suitable for various critical applications like sensing, catalysis, electrodes for energy applications etc. Additionally, they also combine properties of metals like good electrical and thermal conductivity and ductility. They can be fabricated using top-down or bottom-up approaches also known as dealloying and templating which give the fabricator room to tailor properties according to need. In addition, they could find potential applications in many relevant fields in current scenario like drug delivery vehicles. However, there is a long way to go to extract its whole potential.

Keywords: metallic nanostructures, nanoporous metallic films, nano fabrication, optical sensing, catalysis

1. Introduction

Nanoscience is the study of phenomena and manipulation of materials at atomic, molecular and macromolecular scales (1 Bohr radius = 0.5292 Å ≈ 0.05 nm). Due to the influence of the negligible dimensions, materials exhibit remarkable functionality and phenomena. Here properties differ significantly from those at larger scale, because at this level, quantum mechanics and statistical mechanics come into play instead of classical mechanics, and the extremely high surface area to volume ratio of the particles modifies the electrical and chemical activity of the substance, thus the effective concentration of reactants confined in nanostructures may be very high. Typical nano-systems may contain from hundreds to tens of thousands of atoms.

Presently scientists and engineers are finding a wide variety of ways to deliberately make materials at the nano-scale to take advantage of their enhanced properties such as higher strength, lighter weight, increased control of light spectrum, and greater chemical reactivity than their larger-scale counterparts [1], and these products have various applications and a niche market in the fields of electronics, chemistry and biomedicine; semiconductor technology being of the most significance. Needless to say, this branch of science has a great of scope for research and innovation for years to come, on its way to increasing process efficiency, cost effectiveness, and broadening the range and accuracy of human perception.

Out of all the nanostructures, nanoporous film has attracted many research groups in the past decade due to the presence of nano holes in it which acts as a nanoparticle and increases the specific surface area. Also, the increase in chemical stability plays a role in the attraction of several research groups. In order to investigate it further various metals have been used to fabricate nanoporous metallic films. Nanoporous Ag cathode has been used photoelectrochemical carbon dioxide

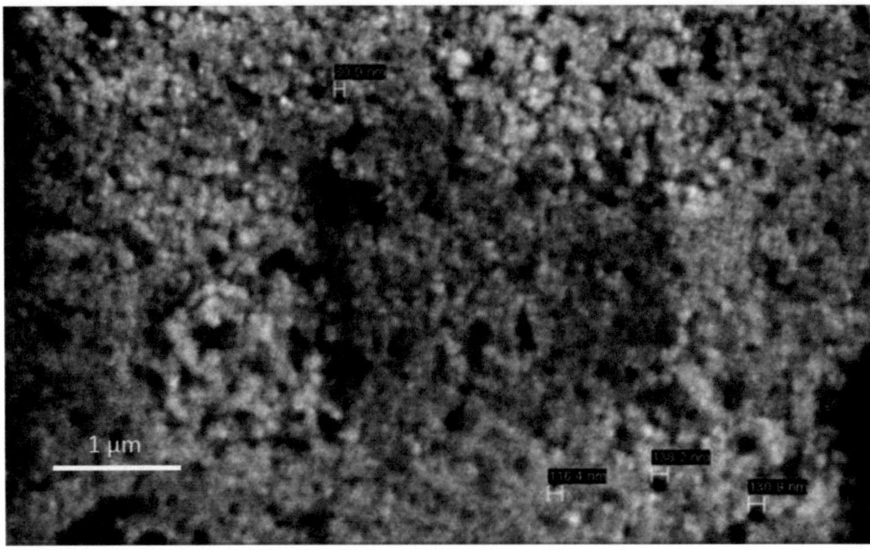

Figure 1.
SEM image: Electroless plated Au on e-beam evaporated Cu on silicon substrate [5].

reduction [2] and nanoporous palladium have been used for reductive dichlorination [3]. But among all the metallic porous films, the nanoporous gold (NPG) film is used mostly due to its chemical stability and unique surface chemistry.

NPG film provides a suitable microenvironment for immobilization of biomolecules (like enzymes) by maintaining their biological activity and facilitates electron transfer between the immobilized proteins and electrode surfaces, leading to its intensive usage in electrochemical biosensors with enhanced analytical performance compared to other biosensor designs [4].

Figure 1 shows surface morphology of NPG film which was manufactured by e-beam deposition of Cu on silicon substrate and then electroless plated with Au. The characteristics of Au nanoporous film such as high surface-to-volume ratio, high surface energy, ability to decrease proteins metal inter-particulate distance, and the functioning as electron conducting pathway between prosthetic groups and the electrode surface have been claimed as the reasons to facilitate electron transfer between redox proteins and electrode surfaces. And these are the properties that make the NPG film so coveted for the fabrication of electrochemical sensors and biosensors.

In this chapter, the processes to make nanoporous metallic films followed by effect on properties of the film are discussed. The applications of nanoporous metallic films and finally, the future scope are also elucidated.

2. Fabrication methods

The fabrication techniques for the preparation of nanoporous structures vary with the requirement of the application. The nanoporous films can be either etched out from an alloy (top-down approach), known as dealloying process or can be fabricated using a template and removing it after deposition of required metal on it (bottom-up approach), known as a templating process [6]. The stability of a nanoporous film is dictated by their pore and ligament sizes.

In dealloying, the undesirable material of the alloy is dissolved under appropriate corrosive condition leaving a stable porous desirable metal [7]. Dealloying can also be further divided into free corrosion dealloying and dealloying by using electrochemical methods. The key factor for choosing an alloy for dealloying by free corrosion is the parting limit which suggests that the composition of undesirable material should be higher than a threshold value. The parting limit for $Au_x Ag_{x-1}$ found by experiments done by Newman [8] is 0.4 for x . Selective dealloying has been used by various research groups to fabricate porous structures below 100 nm. The formation of pores during the dealloying process can be divided into three stages according to SAXS (small-angle X-ray scattering) analysis [9]. The time of each stage is inversely proportional to the dealloying temperature. In the first stage, there are some changes in the SAXS data but physically the commencement of pore formation cannot be detected. The second stage shows some physical changes with the increase in visibility of pits on the surface, subsequently, there is a drastic change in the intensity of the SAXS curve. The third stage corresponds to the growth of pores and a visible increase of ligaments in the structure until a stable structure is achieved. When strictly focusing on NPG films of some suitable alloys like AuAg [10], AuCu [11] and AuNi [12], the most commonly used is AuAg alloy.

Introduction of electric potential to facilitate the dealloying process have also been attempted and is known as dealloying by using an electrochemical method. The major component of this process is an anode (the alloy), cathode, reference electrode and electrolyte. Two common electrolytes for AuAg alloy are aqueous perchloric acid ($HClO_4$) for relatively big pore size and neutral silver nitrate solution (AgNO3) for small pore size [13]. This method requires a more sophisticated set-up than free dealloying process but in return, more uniform nanoporous film and a higher degree of process control are achieved. Similar to parting limit, this method also requires a positive potential value known as critical potential (E_c). The selective dissolution takes place by the rapid increase of Ag dissolution rate. As would be explained later in this chapter, any factor which enhances the surface diffusivity would have an impact on pore as well as ligament size.

Dealloying by the electrochemical process has also been divided into two types, potentiostatic dealloying and galvanostatic dealloying. The experimental set-up for both potentiostatic and galvanostatic methods is the same. In potentiostatic dealloying, the potential value is kept just above the critical potential (E_c) value, which facilitates the gradual dissolution of Ag giving a robust and uniform structure at the end. Whereas in galvanostatic dealloying, the potential value starts above E_c and gradually increased till a maximum limit which is also known as cut-off potential. The two competing factors of Ag dissolution which increases the stress and Au diffusion which decreases the stress in NPG film contribute to the quality of pores in a NPG film [13]. Galvanostatic dealloying by controlling Ag dissolution rate and Au diffusion rate through a periodic increase in potential provides a more robust and crack-free NPG film when compared to that of potentiostatic dealloying.

Low pore size and high ligament size related to the high thermal and electrical conductivities is reported by various research groups [14, 15]. For a 1.3 μm thick NPG film, the ligament size can range from 22 to 155 nm [16]. Hakamada [17] while fabricating nanoporous Ni, Ni-Cu and Cu found an inverse correlation between the atomic ratio of Ni in alloy and ligament size. Another important factor that determines the ligament size is surface diffusion at the metal/electrolyte interface. Correlation between surface diffusion coefficient (D_s) and ligament size (d) is given by Equation [18];

$$D_s = \frac{d^4 kT}{32\,\gamma t a^4} \tag{1}$$

where k is Boltzmann constant, T is the absolute temperature, γ is the surface energy, t is the dealloying time and a is lattice parameter. According to the above relationship, surface diffusion also depends on dealloying temperature and time. Qian and Chen [18] quantified the temperature dependence of NPG films by increasing temperature from −20 °C to 25 °C leading to an increase in diffusivity by two orders of magnitude. Apart from the above-mentioned factors, the dealloying process gets affected by properties of precursor alloy and dealloying solution [19, 20].

The disadvantages of dealloying is the effect of acids and bases used as a solution on the workforce as well as wastage of the dissolved metal. To cope with this problem, Zhang [21] have used ultrasonic irradiation (UI) to assist the dealloying process. This additional method uses lower acid concentration and simultaneously reduce environmental pollution. The UI reduces the surface energy which further enhances the diffusion leading to more coarsening of the ligaments [22]. This experiment proved that the coarsening rate of the ligament increases by introducing UI in dealloying process.

Similar to the use of ultrasonic irradiation, ultrasonic agitation has been used to achieve finer ligaments and pores of palladium-nickel nanoporous thin films [23]. The ultrasonic agitation reduced the time by a factor of 5 without disturbing the desired structure. There has been a similar effect of the magnetic field highlighted on the nanopores of Ag [24].

Another method for the fabrication of NPG films is the templating process. The templating process can be explained in two steps, the preparation of Au or Ag-Au coated core/shell particles followed by the removal of core material to get pure metal foam [25]. The preferred material for template assisted fabrication of NPG film is silica or polystyrene beads. This method gives a higher level of control over pore and ligament size as these would be dependent on the size of beads that can be readily controlled during template fabrication.

3. Properties of nanoporous gold films

3.1 Mechanical properties

Using the analogy of foam to describe nanoporous materials, mechanical properties of foam depends on the cell size similarly pore size dictates the mechanical behaviour of nanoporous materials. Though there is a resemblance between both structures, the effect of scale cannot be neglected and the equation of foams for mechanical behaviour cannot be applied. Also, the introduction of capillary actions and the plastic behaviour of ligaments cannot be unforeseen at lower dimensions. Hodge et al. [26] attempted to present an equation from experimental data for yield strength and it should be emphasized that as the ligament size approaches 1.0 μm the data begin to approach the Gibson and Ashby scaling prediction.

$$\sigma^{*} = C_{s}\left[\sigma_{o} + kL^{-\frac{1}{2}}\right]\left(\frac{\rho^{*}}{\rho_{s}}\right)^{3/2} \qquad (2)$$

Where * denotes foam properties and s denotes solid properties, C_s is a fitting coefficient, σ_o is the bulk material yield strength (σ_s), k is the Hall–Petch-type coefficient for the theoretical yield strength of Au in the regime, $\rho * / \rho_s$ is the ratio between densities of the porous structure and corresponding dense material and L is the ligament size. The real picture of what is happening at the nanoscale can only be found by experimenting, so experimental results of yield strength and tensile strength. On the experiment front, the results from pillar compression tests revealed that the yield strength comes closer to theoretical yield strength of Au when the size of pillars decreases. The tensile test on NPG revealed some macroscale brittleness in it which is opposite to the inherent ductile behaviour of Au [13]. This contradiction in behaviour has been checked through another test known as fracture toughness.

Another mechanical property which is of importance is fracture toughness. It was found that fracture toughness of NPG is low even though gold is inherently ductile [27]. But when the previous phenomena of a tensile test revealing the macroscale brittleness are combined with the above results, the contradictory behaviour becomes clear. In nanoporous films, the ligament acts as a pillar to support the structure. So, the combined behaviour of the structure is coming from the intrinsic behaviour of ligaments. Li and Sieradzki [28], also correlated the ligament size and fracture behaviour. Research groups also concluded the rupturing of ligaments below 100 nm in size [29, 30]. The change in facture behaviour has been observed when the amount of Ag was varied in the final product. For less than 1% of Ag in final nanoporous structure, the rupture is smoother than the other increased value of Ag suggesting the rupturing is intragranular for lesser Ag content [13]. This observation means for less than 1% Ag, the grain boundary strength is higher or the whole system is more brittle as it broke without showing a significant change in appearance. But when the amount of Ag is increased the crack propagates through grain boundaries. Though there is significant data for these behaviours, intensive research is required to fully understand the phenomena.

3.2 Optical properties

The optical properties of metals are dictated by the to and fro motion of the electrons in the outer shell of metal that are triggered by any electromagnetic radiation. The motion can simply be understood by imagining photoelectric effect. The surface electrons are known as surface plasmons (SPs). The variables in this phenomenon are metal since each metal releases a unique amount of energy which acts as the fingerprint of that metal and frequency of electromagnetic radiations. Therefore, by changing these variables a nanoporous structure can be used for a huge number of applications like sensors [31], medical imaging, diagnostics [32] etc. Based on the movement of surface plasmons (SPs), the optical characterization techniques have been classified into surface plasmon polaritons (SPPs) and localized SPRs (LSPRs). With the help of excitation from grating or prism couplers, SPPs are known to propagate for tens or hundreds of micrometers [33]. As the name suggests the second one, localized SPRs (LSPRs) are non-propagating type and since the resonance in a confined space has been associated with a strong electromagnetic field, LSPRs contribute to several significant phenomena like surface-enhanced Raman spectroscopy [34], phononic effects [31]. This strong electromagnetic field becomes more prominent when the nanostructures have sharp features.

So, an ideal nanostructure would be the one which supports both localized as well as propagating systems. The simultaneous presence of a planar structure and nanostructure makes nanoporous materials an ideal candidate with good optical properties. This bicontinuous structure facilitates high field enhancements and good directional control [13]. The relation between irradiation wavelength and propagation of NPR has also been established. The longer the laser wavelength, the farther the propagating SPRs [35].

4. Applications of nanoporous gold films

The nanoporous materials field has gained much attention from the industry due to its enormous specific surface area, well-defined pore sizes and functional sites [36]. Surely, these properties can be achieved for other nanostructures too, but the low capital, high throughput and ease of control of morphology involved in the manufacturing make nanoporous materials more attractive. Among all the metals used for nanoporous structures, Au stands as an outstanding material due to its high surface area (~10 m^2/g), electrochemical activity, biocompatibility and ease of preparation [37, 38]. Due to the enormous surface to volume ratio of NPG, they have shown exceptional sensitivity and selectivity [39]. Particularly sensitivity becomes very crucial in medical or manufacturing safety field, concerning the placement of sensors on which sometimes many lives depend. This is the reason; NPG is finding its way into medical and manufacturing safety field more rapidly.

4.1 Optical sensing

As described in the previous section, the generation of surface plasmon resonance is due to the reduction of the dimension of metals to the scale of the mean free path of electrons [40]. When the electromagnetic radiations of the surroundings interact with electrons, there is inelastic scattering which depends on the pore size. In general, the smaller the pore size the higher the sensitivity [39].

Lang [41] studied the effect of varying nanoporosity on the enhancement of fluorescence. A new method of fabrication was introduced using a combination of dealloying and electroless plating to fabricate NPG structure with high ligament size. This enlarged ligament size facilitated the weakening of plasmon dampening leading to the enhancement in surface-enhanced fluorescence. It was further reported by Lang et al. [42] that fluorescent intensity of molecules absorbed on human-serum-albumin (HSA)-coated NPG films is inversely proportional to the nanopore size. The 45 times increase of fluorescence intensity was reported for a pore size of ~10 nm using this method. Whereas Zhang [43] fabricated a NPG film based optical sensor for sub-ppt detection of mercury ions. A Cy5-labelled aptamer NPG sensor was used with resonant excitation laser, to achieve 0.2 ppt Hg^{2+} sensitivity. This sensor could be extended further for detection of other heavy metal ions.

Similarly, there have been many studies on the surface-enhanced Raman spectroscopy. Zhang [44] have modified the nanoporous structures with wrinkles to include more "hot spots" for ultrahigh SERS enhancements. This was achieved with the help of thermal contractions of prestrained polystyrene microparticles (PS). The wrinkled NPG was found to have 100 times higher signal than the normal NPG films. Another interesting optical application was reported by Shih [5] where they have used NPG gold disks to sense chemical and find refractive index simultaneously. The NPG disks modified with octadecanethiol (ODT) and the surface-enhanced near-infrared absorption (SENIRA) spectroscopy was used to detect hydrocarbon compounds from crude oil samples.

Figure 2 shows the enhancement in SERS spectra when Au nanostructures are formed Pt substrates rather than on Cu substrate [45]. This research work also proves the importance of selection of substrate for use in optical sensing phenomena.

4.2 Electrochemical sensing

Electrochemical sensors are an electrode which goes through a redox reaction to detect the substance attached to the sensors. Now, the sensitivity and selectivity of the sensor become the prominent property to tune for respective applications. A schematic diagram of the sensor has been shown in **Figure 3**.

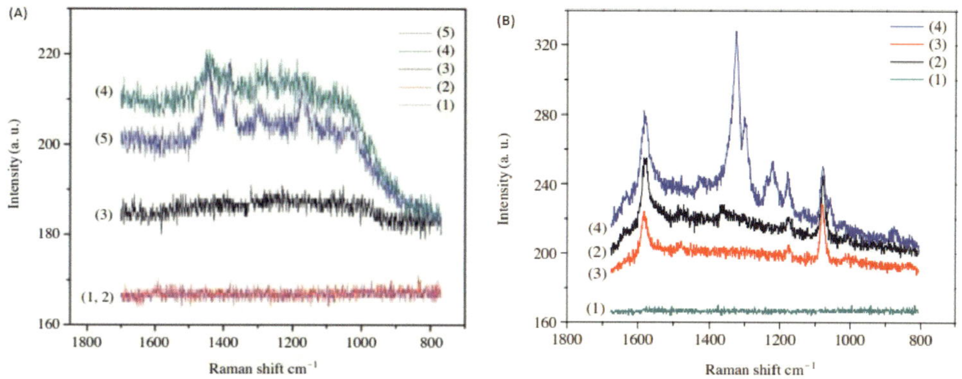

Figure 2.
(A) SERS spectra constituting SERS signal from (1) bare e-beam Au sample and Cu substrate whose reaction times are (2) 0 min, (3) 2 min, (4) 18 h and (5) 24 h respectively. (B) SERS spectra obtained from Pt substrate whose reaction times are (1) 0 min, (2) 2 min, (3) 19 h and (4) 24 h respectively [45].

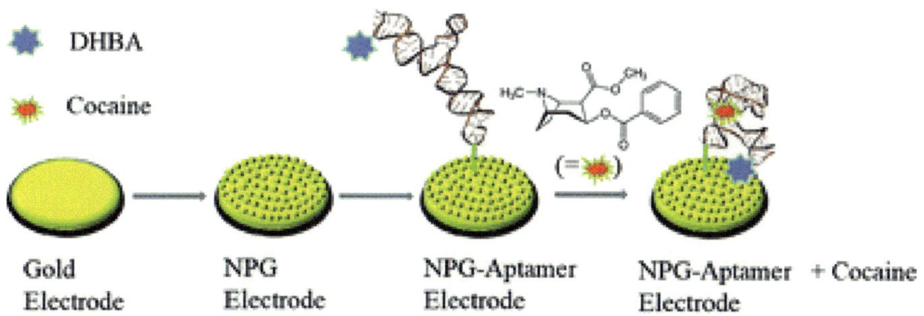

Figure 3.
Working of a NPG electrochemical sensor [46].

Electrochemical sensors are being used in biomedical applications on a large scale due to its sensitivity and selectivity. Chen [47] fabricated an electrochemical NPG film sensor to detect glucose based on the current response. As it was observed in optical sensors, lower pore size gave better sensitivity for glucose in this electrochemical sensor. The sensor was fabricated with the help of dealloying method and then cyclic voltammogram (CV) curves for NPG was used to detect OH^- adsorptions as it has direct correlations with electro-oxidation of glucose. In order to check the selectivity and sensitivity towards glucose of the sensor, glucose concentration was varied by keeping the pore size (18 nm) and current potential constant (0.1 and 0.3 V). On comparison, the current density at 0.1 V decreased while it increased linearly for 0.3 V proving the sensitivity of electrochemical sensor towards oxidation and subsequently towards the concentration of glucose. Additionally, the sensor was evident of excellent selectivity by avoiding interference caused by other substance present in the solution.

Electrochemical NPG sensor was used by a research group for the detection of DNA [48]. The biosensor showed an excellent sensitivity with a limit of detection up to 28 aM. The fact that the nanopores capture DNA and immobilizes it makes it more selective. Likewise, simple fabrication technique of dealloying makes it more feasible. Qui [49] went one step further to enzyme-modify NPG electrochemical biosensors to detect glucose and ethanol. The NPG was modified with the help of alcohol dehydrogenase (ADH) or glucose oxidase (GOD) that enhanced its sensitivity towards glucose and ethanol. The promising fact about these sensors is even after leaving them for 1-month storage at 4 ◦ C, the ADH- and GOD- based biosensor lost only 5% and 4% efficiencies, respectively. In this connected world, where some product is manufactured at one place and then transported to another sustained efficiency is of prime importance.

4.3 Catalysis

Catalysis is another activity which is highly dependent on the surface area for its efficiency. The first catalytic activity of gold nanoparticles was reported to be back in 1987 when CO was oxidized far below room temperature [50]. Due to inherent inert behaviour of gold, this experimental result came as a surprise. Moreover, a nanostructure is constructed on a substrate. So, when a reaction was taking place of these nanostructures used to come off from the substrate as a result of poor adhesion. This is where NPG gained its importance in this field for its bicontinuous structure [51]. In case of oxidation, the high surface area acts as an important site for adsorption giving exposure to a higher number of reactant molecules to interact with the surface. Another reason for high oxidation behaviour is the presence of some amount of Ag in NPG films. It is known that Ag bind oxygen and activate them [52].

Shi et al. [53] used NPG functionalized with praseodymium-titania mixed oxide to catalyze water-gas shift reaction. Both electron energy loss spectroscopy (EELS) and flow reactor tests revealed that Pr-TiO_x functionalized NPG is highly active as well as very stable to high temperature such as 180–400°C. This study exhibited the interaction between Au substrate and the oxide deposit which plays a vital role in the dissociation of water. The problem with the use of any nanostructure was decay in catalytic activity with time due to coarsening of the nanostructure. Use of Pr-TiO_x formed a mixed Pr-TiO_x solid solutions which prevented further coarsening of NPG making the catalyst stable to use for a long period. The catalytic activity of Au is also vital in recent reports of hydrogen fuel. Albeit the produced hydrogen contains a small amount of CO that can further deactivate the electrodes [54], a highly sensitive and selective catalyst is required for this purpose. NPG films form potential candidate for such catalytic applications.

Similar to CO oxidation, research has also been started in H_2 oxidation. Qadir et al. reported very low H_2 oxidation activity by bare np-Au [55]. The activity was manipulated by deposition of titania on the catalyst. This exercise also proves that tuning the amount of titania deposit can increase the oxidation activity of the structure.

4.4 Biomolecular sensing

Modified electrodes are being widely employed in modern electrochemistry for electrocatalytic reactions and as electrochemical sensors. Gold electrodes are useful to construct electrochemical sensors because of their chemical inertness. The well-established strategy of a self-assembled monolayer formation for immobilization of compounds onto gold surfaces are based on the attachment of thiol (-SH) or disulfide (-S-S-) functional groups to Au (111) [56].

In order to develop new reliable, efficient and functional micro/nanoscale devices, control over the surface properties is essential. The surface properties of microscale and nanoscale devices can easily be controlled and manipulated in a versatile manner through surface modification technology. The properties such as wetting, biocompatible, bioselective, optical and electronic characteristics of various inorganic and polymeric surfaces can be adjusted and controlled by modifying the surface.

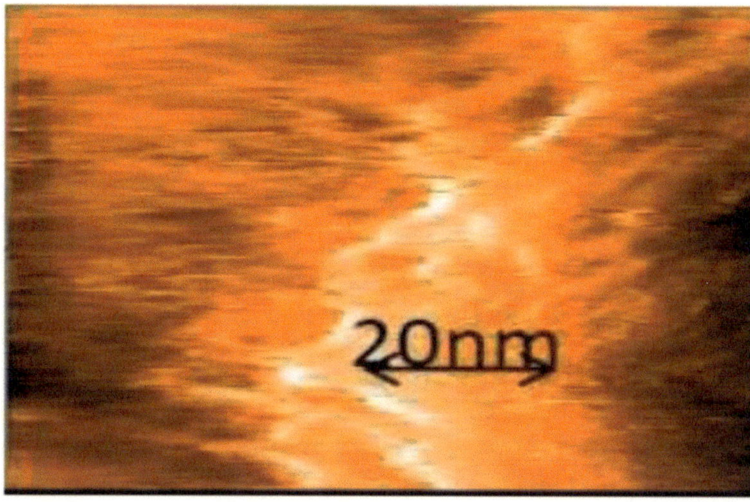

Figure 4.
STM image of gold on mica with surface modification by L-cysteine molecules [57].

Furthermore, a wide variety of terminal functional groups such as amino group, carboxylic acid group can be implemented to detect the trace heavy metal ions, DNA, RNA or antibodies.

Surface modifications can be grouped into two broad categories: (a) Chemically or physically altering the atoms or molecules in the existing surface (treatment, etching, chemical modification) (b) Coating over the existing surface with a material having a new composition (solvent coating or thin film deposition by chemical vapour deposition, radiation grafting, chemical grafting or RF-plasmas).

The Self Assembled Monolayers (SAMs) are nanostructures that are formed by organic assemblies owing to the adsorption of molecular constituents from solution or gas phase onto the surface of solids or arrays on liquid phase. The molecules or ligands that form SAMs have a chemical functionality called "headgroup" which has a special affinity towards a substrate. Typically, the thickness of a SAM is typically 1–3 nm.

SAMs are well-suited for studies in nanoscience and nanotechnology because: They are easy to prepare. They do not require ultrahigh vacuum (UHV) or other specialized equipment. SAMs can be easily prepared by immersing the substrates into the known solution. They form on objects of all sizes and are critical components for stabilizing and adding function to preformed, nanometer-scale objects for example, thin films, nanowires, colloids, and other nanostructures. They can couple the external environment to the molecular, electronic and optical properties of metallic structures. The most extensively studied class of SAMs is derived from the adsorption of alkanethiols (-SH) on gold, silver, copper, palladium, and mercury. SAMs also provide a convenient approach for ultra-low-level analyte recognition and have been important in the development of electroanalytical devices and electrochemical sensors [56].

Above **Figure 4** shows the surface coverage of L-Cysteine molecules on NPGF and with the increased surface coverage L-Cysteine molecules would be able to trap more heavy metal ions leading to lesser limit of detection (LOD) [57].

5. Future of NPG

The future for NPG films is promising though there are many unanswered questions. Like the understanding of relationship between constituent of an alloy and its morphology after dealloying and pressure flow relationship NPG sieves as well as

Figure 5.
SEM image of nanoflower [45].

its membrane architecture [58]. Also, the degree of sensitivity is going to play a vital role in coming days which will strengthen its foothold in the sophisticated sensing applications. The research work is already underway for different structures such as NPG leaves, nanowires, nanoflowers (**Figure 5**), etc. Looking at the promising optical and mechanical properties there is a long way to go before the full potential of NPG is reached. Additionally, with the onset of the decade where data is going to be of so much importance, correct data and large amount of data would be of high importance for precise decision-making purposes.

With the passage of time, the resources are becoming scarce triggering the requirement of tools which utilizes fewer materials to give more information. Also, selectivity would be of prime importance. For increased selectivity the sensors or catalysts should be manipulated from the bottom and this can be possible through NPG film like structures only. The use of minimal material would ensure less environmental impact. Hereby, the research should be more focused in areas like hydrogen fuel which is environmentally friendly and can pave way for potential applications in transportation industry.

6. Conclusion

Nanoporous metals is known to exhibit strong electrochemical, optical and mechanical properties due to their unique three-dimensional and quasi-periodic nanoporosity. Nevertheless, there are many challenges that remain. Performance of electrochemical sensors in energy applications depends strongly on their structure and composition [59]. So, new electrochemical fabrication method with the ability of tailoring of the size and shape of nanomaterials is required. Similarly, understanding of nanoporous metals with improved optical performances is necessary as it needs superior reproducibility, facile synthesis and excellent stability [13]. Additionally, in the field of medical research nanoporous metals are can be used for controlled drug-delivery [60]. Nanoporous metallic films would be attractive materials for future applications research that would result in huge advancement of the field of technology with strong conviction.

Acknowledgements

The authors would like to thank BITS Pilani OPERA Award and Reliance Industries Limited Education assistantship.

Thanks/other declarations

The authors do not have any difference in opinion. There is no conflict of interest.

Author details

Swastic[1] and Jegatha Nambi Krishnan[2]*

1 University of Leeds, Leeds, UK

2 BITS Pilani K K Birla Goa Campus, Goa, India,

*Address all correspondence to: jegathak@goa.bits-pilani.ac.in

References

[1] Chong KP. Nano science and engineering in solid mechanics. Acta Mechanica Solida Sinica. 2008 Apr 1;21(2):95-103.

[2] Zhang Y, Luc W, Hutchings GS, Jiao F. Photoelectrochemical carbon dioxide reduction using a nanoporous Ag cathode. ACS applied materials & interfaces. 2016 Sep 21;8(37):24652-24658.

[3] Li W, Ma H, Huang L, Ding Y. Well-defined nanoporous palladium for electrochemical reductive dechlorination. Physical Chemistry Chemical Physics. 2011;13(13):5565-5568.

[4] Nguyen HH, Lee SH, Lee UJ, Fermin CD, Kim M. Immobilized enzymes in biosensor applications. Materials. 2019 Jan;12(1):121.

[5] Shih WC, Santos GM, Zhao F, Zenasni O, Arnob MM. Simultaneous chemical and refractive index sensing in the 1-2.5 μm near-infrared wavelength range on nanoporous gold disks. Nano letters. 2016 Jul 13;16(7):4641-4647.

[6] Jang SW, Hwang S, Lim SH, Han S. Fabrication of nanoporous thin films via radio-frequency magnetron sputtering and O2 plasma ashing. Vacuum. 2019 May 1;163:81-87.

[7] Lee G-H, An S, Jang SW, Hwang S, Lim SH, Han S. Fabrication of nanoporous noble metal thin films by O 2 plasma dealloying. Thin Solid Films. 2017;631:147-151.

[8] Newman RC, Corcoran SG, Erlebacher J, Aziz MJ, Sieradzki K. Alloy corrosion. MRS Bulletin-Materials Research Society. 1999 Jul 1;24:24-28.

[9] Lin B, Kong L, Hodgson PD, Mudie S, Hawley A, Dumée LF. Controlled porosity and pore size of nano-porous gold by thermally assisted chemical dealloying–a SAXS study. RSC advances. 2017;7(18):10821-10830.

[10] Newman RC, Sieradzki K. Metallic corrosion. Science. 1994 Mar 25;263(5154):1708-1710.

[11] Morrish R, Dorame K, Muscat AJ. Formation of nanoporous Au by dealloying AuCu thin films in HNO3. Scripta Materialia. 2011 May 1;64(9):856-859.

[12] Rouya E, Reed ML, Kelly RG, Bart-Smith H, Begley M, Zangari G. Synthesis of nanoporous gold structures via dealloying of electroplated Au-Ni alloy films. ECS Transactions. 2007 Sep 28;6(11):41.

[13] Newman R, Sieradzki K, Renner FU, Hodge A, Balk J, Kysar JW, Okman O, Ding Y, Ma H, Weissmüller J, Shao L. Nanoporous gold: from an ancient technology to a high-tech material. Royal Society of Chemistry; 2012 Mar 28.

[14] Yu J, Ding Y, Xu C, Inoue A, Sakurai T, Chen M. Nanoporous metals by dealloying multicomponent metallic glasses. Chemistry of Materials. 2008 Jul 22;20(14):4548-4550.

[15] Yang Q, Liang S, Han B, Wang J, Mao R. Preparation and properties of enhanced bulk nanoporous coppers. Materials Letters. 2012 Apr 15;73:136-138.

[16] Gwak EJ, Kang NR, Baek UB, Lee HM, Nahm SH, Kim JY. Microstructure evolution in nanoporous gold thin films made from sputter-deposited precursors. Scripta Materialia. 2013 Nov 1;69(10):720-723.

[17] Hakamada M, Mabuchi M. Preparation of nanoporous Ni and Ni–Cu by dealloying of rolled Ni–Mn

and Ni–Cu–Mn alloys. Journal of Alloys and Compounds. 2009 Oct 19;485(1-2):583-587.

[18] Qian LH, Chen MW. Ultrafine nanoporous gold by low-temperature dealloying and kinetics of nanopore formation. Applied Physics Letters. 2007 Aug 20;91(8):083105.

[19] Dan Z, Qin F, Yamaura SI, Sugawara Y, Muto I, Hara N. Dealloying behavior of amorphous binary Ti–Cu alloys in hydrofluoric acid solutions at various temperatures. Journal of alloys and compounds. 2013 Dec 25;581:567-572.

[20] Hakamada M, Takahashi M, Furukawa T, Tajima K, Yoshimura K, Chino Y, Mabuchi M. Electrochemical stability of self-assembled monolayers on nanoporous Au. Physical Chemistry Chemical Physics. 2011;13(26):12277-12284.

[21] Zhang R, Wang X, Huang JC, Li F, Zhang Z, Wu M. Formation mechanism of nanoporous silver during dealloying with ultrasonic irradiation. RSC advances. 2019;9(18):9937-9945.

[22] Kim MS, Nishikawa H. Fabrication of nanoporous silver and microstructural change during dealloying of melt-spun Al–20 at.% Ag in hydrochloric acid. Journal of Materials Science. 2013 Aug 1;48(16):5645-5652.

[23] Li WC, Balk TJ. Achieving finer pores and ligaments in nanoporous palladium–nickel thin films. Scripta Materialia. 2010 Feb 1;62(3):167-169.

[24] Song T, Gao Y, Zhang Z, Zhai Q. Influence of magnetic field on dealloying of Al-25Ag alloy and formation of nanoporous Ag. CrystEngComm. 2012;14(10):3694-3701.

[25] Biener J, Nyce GW, Hodge AM, Biener MM, Hamza AV, Maier SA. Nanoporous plasmonic metamaterials. Advanced Materials. 2008 Mar 18;20(6):1211-1217.

[26] Hodge AM, Biener J, Hayes JR, Bythrow PM, Volkert CA, Hamza AV. Scaling equation for yield strength of nanoporous open-cell foams. Acta Materialia. 2007 Feb 1;55(4):1343-1349.

[27] T. J. Balk, N. Briot, D. S. Gianola, T. Kennerknecht and C. Eberl, unpublished, 2010

[28] Li R, Sieradzki K. Ductile-brittle transition in random porous Au. Physical Review Letters. 1992 Feb 24;68(8):1168.

[29] Biener J, Hodge AM, Hamza AV. Microscopic failure behavior of nanoporous gold. Applied Physics Letters. 2005 Sep 19;87(12):121908.

[30] Balk TJ, Eberl C, Sun Y, Hemker KJ, Gianola DS. Tensile and compressive microspecimen testing of bulk nanoporous gold. Jom. 2009 Dec 1;61(12):26.

[31] Xiao Y, Xu D, Medina FJ, Wang S, Hao Q. Thermal studies of nanoporous thin films with added periodic nanopores—a new approach to evaluate the importance of phononic effects. Materials Today Physics. 2020 Jan 24:100179.

[32] Jain PK, Huang X, El Sayed IH, El-Sayed MA. Noble metals on the nanoscale: optical and photothermal properties and some applications in imaging, sensing, biology, and medicine. Accounts of chemical research. 2008 Dec 16;41(12):1578-1586.

[33] Homola J. Surface plasmon resonance sensors for detection of chemical and biological species. Chemical reviews. 2008 Feb 13;108(2):462-493.

[34] Stewart ME, Anderton CR, Thompson LB, Maria J, Gray SK, Rogers JA, Nuzzo RG. Nanostructured plasmonic sensors. Chemical reviews. 2008 Feb 13;108(2):494-521.

[35] H. Raether, Surface Plasmons on Smooth and Rough Surfaces and on Gratings, Springer, Berlin, 1986

[36] van der Zalm J, Chen S, Huang W, Chen A. Recent Advances in the Development of Nanoporous Au for Sensing Applications. Journal of The Electrochemical Society. 2020 Jan 10;167(3):037532.

[37] Yan M. Development of New Catalytic Performance of Nanoporous Metals for Organic Reactions. Springer Science & Business Media; 2014 Mar 24.

[38] Webster TJ, editor. Nanomedicine: Technologies and applications. Elsevier; 2012 Oct 19.

[39] Ruffino F, Grimaldi MG. Nanoporous Gold-Based Sensing. Coatings. 2020 Sep;10(9):899.

[40] Pelton M, Bryant GW. Introduction to metal-nanoparticle plasmonics. John Wiley & Sons; 2013 Apr 9.

[41] Lang XY, Guan PF, Fujita T, Chen MW. Tailored nanoporous gold for ultrahigh fluorescence enhancement. Physical Chemistry Chemical Physics. 2011;13(9):3795-3799.

[42] Lang XY, Guan PF, Zhang L, Fujita T, Chen MW. Size dependence of molecular fluorescence enhancement of nanoporous gold. Applied Physics Letters. 2010 Feb 15;96(7):073701.

[43] Zhang L, Chang H, Hirata A, Wu H, Xue QK, Chen M. Nanoporous gold based optical sensor for sub-ppt detection of mercury ions. ACS nano. 2013 May 28;7(5):4595-4600.

[44] Zhang L, Lang X, Hirata A, Chen M. Wrinkled nanoporous gold films with ultrahigh surface-enhanced Raman scattering enhancement. ACS nano. 2011 Jun 28;5(6):4407-4413.

[45] Krishnan JN, Kim IT, Ahn SH, Kim ZH, Cho SH, Kim SK. Electroless deposition of SERS active Au-nanostructures on variety of metallic substrates. BioChip Journal. 2013 Dec 1;7(4):375-385.

[46] Tavakkoli N, Soltani N, Mohammadi F. A nanoporous gold-based electrochemical aptasensor for sensitive detection of cocaine. RSC advances. 2019;9(25):14296-14301.

[47] Chen LY, Lang XY, Fujita T, Chen MW. Nanoporous gold for enzyme-free electrochemical glucose sensors. Scripta Materialia. 2011 Jun 1;65(1):17-20.

[48] Hu K, Lan D, Li X, Zhang S. Electrochemical DNA biosensor based on nanoporous gold electrode and multifunctional encoded DNA– Au bio bar codes. Analytical chemistry. 2008 Dec 1;80(23):9124-9130.

[49] Qiu H, Xue L, Ji G, Zhou G, Huang X, Qu Y, Gao P. Enzyme-modified nanoporous gold-based electrochemical biosensors. Biosensors and Bioelectronics. 2009 Jun 15;24(10):3014-3018.

[50] Haruta M, Kobayashi T, Sano H, Yamada N. Novel gold catalysts for the oxidation of carbon monoxide at a temperature far below 0 C. Chemistry Letters. 1987 Feb 5;16(2):405-408.

[51] Kim SH. Nanoporous gold: Preparation and applications to catalysis and sensors. Current Applied Physics. 2018 Jul 1;18(7):810-818.

[52] Fajín JL, Cordeiro MN, Gomes JR. On the theoretical understanding of the unexpected O 2

activation by nanoporous gold. Chemical Communications. 2011;47(29):8403-8405.

[53] Shi J, Wittstock A, Mahr C, Murshed MM, Gesing TM, Rosenauer A, Bäumer M. Nanoporous gold functionalized with praseodymia–titania mixed oxides as a stable catalyst for the water–gas shift reaction. Physical Chemistry Chemical Physics. 2019;21(6):3278-3286.

[54] Kim SH. Nanoporous gold: Preparation and applications to catalysis and sensors. Current Applied Physics. 2018 Jul 1;18(7):810-818.

[55] Qadir K, Quynh BT, Lee H, Moon SY, Kim SH, Park JY. Tailoring metal–oxide interfaces of inverse catalysts of TiO 2/nanoporous-Au under hydrogen oxidation. Chemical Communications. 2015;51(47):9620-9623.

[56] Love JC, Estroff LA, Kriebel JK, Nuzzo RG, Whitesides GM. Self-assembled monolayers of thiolates on metals as a form of nanotechnology. Chemical reviews. 2005 Apr 13;105(4):1103-1170.

[57] Deshpande A., Joshi M., Krishnan J.N., Priydarshi K., Ramanan S.R., Swastic S., Vemula J.K. Nanocharacterization studies on Surface Modified Nanoporous Gold Films for Sensor Applications. TechConnect Briefs. 2016;1:5-8.

[58] Seker E, Reed ML, Begley MR. Nanoporous gold: fabrication, characterization, and applications. Materials. 2009 Dec;2(4):2188-2215.

[59] Abdel-Karim R. Nanoporous Metallic Foams for Energy Applications: Electrochemical Approaches for Synthesizing and Characterization. Handbook of Nanomaterials and Nanocomposites for Energy and Environmental Applications. 2020:1-24.

[60] Reed ML, Lye WK. Microsystems for drug and gene delivery. Proceedings of the IEEE. 2004 Nov 8;92(1):56-75.

Chapter 8

Nanoporous Carbon Materials toward Phenolic Compounds Adsorption

Mahmoud Fathy Mubarak, Alshimaa Maher Ahmed and Sahar saad Gabr

Abstract

Nanoporous carbon-based sorbents are used to generate a three-dimensional real-space model of the nanoporous structure using the concept of Gaussian random fields. This pore model is used to derive important pore size characteristics, which are cross-validated against the corresponding values from gas sorption analysis. After filling the model pore structure with an aqueous electrolyte and rearranging the ions *via* a Monte Carlo simulation for different applied adsorption potentials. In comparison to nanopores formed from solid-state membranes (e.g., silicon oxide, aluminum oxide, polymer membranes, glass, hafnium oxide, gold, etc.) and very recently 2D materials (e.g., boron nitride, molybdenum disulfide, etc.), those nanopores produced from carbon materials (e.g., graphene, carbon nanotubes (CNTs), diamond, etc.), especially those from graphene appear to be perfect for adsorption process. The thickness of carbon structures nanopores can be as thin as 0.35 nm, resembling the height of the base spacing. Moreover, the sizes of carbon structures nanopores can be precisely fabricated and tuned to around 1.0 nm, the similar size of many heavy metals and organic pollutants molecules. Furthermore, carbon materials are chemically stable and feature-rich surface chemistry. Therefore, various carbon nanopore sequencing techniques have been developed. Finally, in this chapter the adsorption of phenolic compounds on nanoporous carbon specifically the active carbon are overviewed and how to affect the heterogeneity of activated carbon surface, PH of the solution on the efficiency of adsorption.

Keywords: nanoporous carbon, phenolic compounds, adsorption, activation process, templating methods

1. Introduction

Contamination of water is one of the significant issues in the universe, that poses negative effects on individual and surroundings. The rising in industrial and human activities resulted in increasing the flowing of wastewater into water supplies [1, 2]. In the last years, the impacts of exposure of human and animals to chemicals in the ambiance especially the aquatics medium has taken the high interest of many scientists and decision-makers [3]. Among these chemicals, phenolic compounds are considered the most important due to their toxic effect on animals and humans that result from their staying in the ambient for along time and then collect to cause that effect [3]. There are two types of phenolic compounds, natural compounds connected to the flowers and fruits colors and synthetic compounds used in daily humans life for various purposes [3]. Phenolic compounds are present in the effluents of various industries such as oil refining, petrochemicals, pharmaceuticals, coking operations, resin manufacturing, plastics, paint, pulp, paper, and wood products. Discharge of these compounds without treatment may lead to serious health risks to humans, animals, and aquatic systems [4]. The presence of these compounds is attributed to a breakdown of natural organic materials in the water, flows water away from farmland, and discharge of wastes resulting from industries and humans uses in water resources. The presence of these compounds in water results in the interaction of them with chemical, physical and biological variables inside the water that led to their conversion to other forms that have a dangerous effect than the original ones [3, 5, 6]. Phenol has been designated as a priority pollutant by the US Environmental Protection Agency (EPA) and the National Pollutant Release Inventory (NPRI) of Canada [3]. International regulatory bodies have set strict discharge limits for phenols for a sustainable environment. For example, the EPA has set a water purity standard of less than 1 ppb for phenol in surface water [7]. The toxicity levels usually are in the range of 9–25 mg/L for both humans and aquatic life [3, 8]. Phenolic compounds are categorized as very harmful contaminants due to their toxic effects and cancer diseases causing. Short -term exposure to these compounds results in irritation in human organs, headache, and inability to balance even at low content, while Long-term exposure to these compounds causing arise in the pressure of blood and very strong kidney and liver problems [9, 10].

Phenolic compounds removal from water is necessary to protect humans and aquatics from the pollution that those toxic compounds causing. Appearing a lot of methods used in the phenolic compounds removing will overcome the hazard problems connected to these chemicals and wastes discharge challenges, in addition to, the getting of additional value phenolic compounds as secondary products. A lot of technologies are used to remove phenolic compounds from wastewater successfully before it's disposal in water resources [3, 7].

Electrochemical oxidation [2, 3, 8], (electro)chemical coagulation [10], solvent extraction [3], bioremediation [10], and photocatalytic degradation [3, 8], Reverse osmosis and nanofiltration [2, 8], Chemical oxidation [2, 8], have been used for the treatment of wastewater from phenolic compounds for many years, but these techniques are very costly due to the requirements they needed in purification process as supplementary chemical materials and high input of energy, in addition to the undesired by-products produced through the treatment process. Therefore, the separation of phenolic compounds from wastewater requires the development and using energy-efficient and cheap methods [5, 8, 11–14]. In this Chapter, the adsorption method is very effective for that purpose. Adsorption is the most effective method for removing the organic and inorganic contaminants from wastewater because it is a very easy method to set up, low cost, no time consuming, the adsorbent used in the process not harmful to the environment and can be recovery and reused again without the decrease in the efficiency [10]. In the adsorption method, the removal of pollutants from water occurs by holding them on the adsorbent surface [2, 11–13]. Carbon-based nanomaterials such as fullerenes, carbon nanotubes (CNTs), graphene and its derivative compounds, nanodiamonds, and nanoporous carbons (NPC) such as activated carbon are the most popular nano adsorbent materials used for purification of water between scientists due to their harmless natural to the environment, abundance, simplicity of handling, and size and form that give them different properties [15–19]. In this chapter, nanoporous carbons (NPC) is considered one of the most effective and economical adsorbents used in the separation of organic and inorganic contaminants from the aquatic environment due to their various properties such as high surface area and high porosity, in addition to they are inexpensive, abundance, generate from renewable sources, very thermally stable, and their perfect chemical resistance [20–22]. Furthermore, NPC can attract attention to used in many purposes due to its rule in decreasing the amount wastes in the environment through using them in Their preparation process. Their unique properties offer new opportunities in the area of inclusion chemistry, guest host interaction, and molecular manipulations, showcasing their potential impact in a wide range of research fields, such as adsorption, separation, catalysis, electronic devices, and drug delivery [18, 19, 21].

Activation process(such as physical or thermal activation and chemical activation methods) is one of the methods used to prepare nanoporous carbon but due to the disadvantages of this process, the vision has been directed to using the templating method in the preparation process [19, 23].

1.1 The aim of this chapter

To prepare nanoporous carbon materials (NPC) to use in phenolic compound removal, discussing preparation methods, properties of these materials especially activated carbon, and improving these properties to improve the performance of these materials in adsorption application by using templating methods.

2. Phenolic compounds

Nowadays there is a growing concern around the world constantly about the increasing volume of pollutants in the water and the removal of dangerous pollutants from wastewater is one of the most important environmental issues at present. Phenolic derivatives are among the common environmental pollutants. The extremely low concentration of these pollutants is an obstacle to water use. Phenols are toxic and carcinogenic that can cause a bad taste and smell in drinking water also harmful to human health [5, 24]. Phenolic compounds come to water from different various sources such as oil refineries, coal gasification sites, petrochemical units, and from the synthesis of plastics, paints, pesticides, insecticides, pharmaceutical, etc. according to European Union countries, the maximum concentration of phenols in the drinking water is limited to .5 ppb and in the USA to 1 ppb. Both the US Environmental Protection Agency (EPA) and the European Union (EU) involve indicated that nitrophenols and chlorophenols rank first on the pollutant list. The most widespread of phenolic compounds in water is chlorophenol that generates from the chlorination of aromatic compounds that present in water and soil. Phenols have weak acidic properties. (**Table 1**) discuss the basic information about most phenolic compounds [5, 6, 9].

Diverse technologies have been employed for the removal of phenolic compounds from a variety of water sources including steam distillation, aerobic and anaerobic biodegradation, oxidation by ozone, ion-exchange resins, adsorption, and membrane filtration [11]. But above-mentioned methods, adsorption is the most applied technique for water treatment due to its very simple technique as it works by adding the adsorbent to the polluted water and then target pollutants are adsorbed into the adsorbent, cost-effective, friendly environment and the availability of a wide range of adsorbents. The adsorption of phenol and its derivatives on nanoporous carbon especially activated carbon has become an important issue by many researchs [9, 25]. Adsorption is the most applied technique for the removal

Phenolic compound	pKa at 25 °C	Critical Oxidation Potential (COP)	Boiling point (°C)	Solubility at 25 °C ($g\ l^{-1}$)
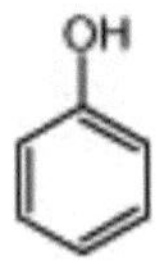	9.89	1.089	182	93
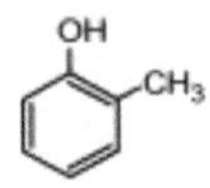	10.20	1.040	191	25
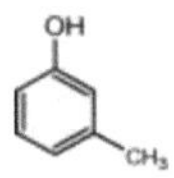	10.01	1.080	202	26
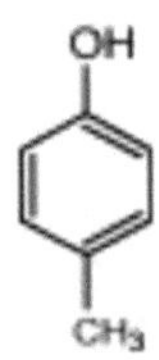	10.17	1.038	202	23
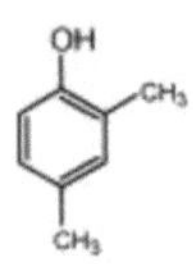	10.58	0.895	211	Unavailable
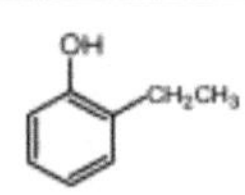	10.2	Unavailable	207	Poorly soluble
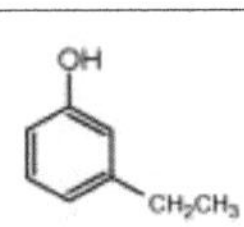	10.07	Unavailable	214	Mildly soluble
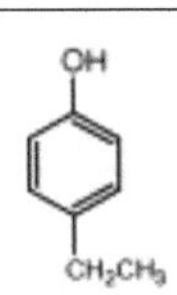	10.0	Unavailable	218	Mildly soluble
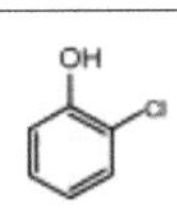	8.52	1.094	204	28
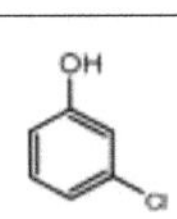	8.97	Unavailable	214	26
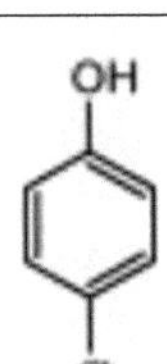	9.37	1.094	220	27

Phenolic compound	pKa at 25 °C	Critical Oxidation Potential (COP)	Boiling point (°C)	Solubility at 25 °C (g l^{-1})
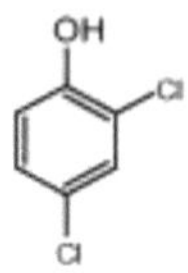	7.90	Unavailable	210	4.5
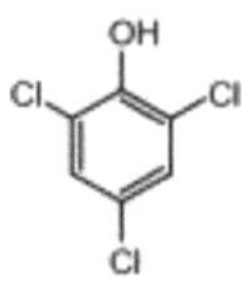	5.99	1.103	246	28.6

Table 1.
The basic information about most phenolic compounds [24].

of phenolic compounds as it is low cost with high efficiency and easy ergonomic design besides that activated carbon is the most applied adsorbent as it has an internal porous structure (consisting of pores of varying size) with large surface area and specific chemical structure of the surface. And the efficiency of adsorption capacity of phenolic compounds on the activated carbon accompanied by numbers of factors such as:

- Nature of surface functionality of adsorbent and its pore structure
- Properties of adsorbate (phenol) such as pKa, functional group, size, and polarity
- Adsorption parameters such as PH of the solution, temperature, dose of adsorbent, and concentration of adsorbate [24, 25].

There are also important factors such as the type of precursors for Activated carbon preparation and the aqueous solubility of phenolic compounds [9, 11, 26].

Despite extensive studies on factors affecting phenol adsorption, the mechanism of its adsorption is unclear and should be further studied. In particular, the most controversial matter is the role of the presence of oxygen group on its surface in the uptake of phenols [24].

In this study, we have tried to explain the importance of the above factors and how to prepare suitable AC from cheap and available precursors to remove phenolic compounds.

3. Historic perspective of nanoporous carbons

Carbon is the most spreading element on the earth, it has distinct characteristics and can form many compounds with different properties. Carbon has been used for a long period in form of coal, charcoal, and carbon black. After that has been discovered a new process to improve the properties of carbon materials through activation of charcoal. These new materials are called nanoporous carbon materials [13, 27, 28].

Carbon was used in past in form of charcoal or carbon black for many purposes:

- It was used as a pigment in the painting of caves from 35000 to 11000 BC.
- Using the carbon in form of charcoal in the production of metal, in 8000 BC.
- Used to decrease minerals in process of bronze fabrication by ancient Egyptians, in 3750 BC
- In medicinal use, since, The first proof was in 1550 BC, where ancient Egyptians used charcoal in the treatment of the stomach tract and removing odors from putrefying injuring.
- In water treatment, since in 450 BC, charcoal filters were used in the treatment of drinkable water [28, 29].
- In the eighteenth century, carbons made from blood, wood, and animals were used for the purification of liquids.
- In the treatment of gases was this observed by the Swedish chemist Karl Wilhelm Scheele in 1773.
- Decolorizing of solutions, experiments on this was performed by Lowitz in 1786.

However, a few years later, in 1794, an English sugar refinery successfully used wood charcoal for decolorization. This application remained a secret until 1812 when the first patent appeared in England, although from 1805 wood charcoal was used in a large-scale sugar refining facility in France for decolorizing syrups, and by 1808 all sugar refineries in Europe were using charcoal as a decolorizer [25, 30].

In 1811, it was proved that the efficiency of decolorization of sugar syrups by bone char was higher than wood char. In 1815, the majority of sugar refining facilities were using granular bone-derived char.

In 1817 Joseph de Cavaillon patented a method of regenerating used bone chars, but the method was not entirely successful [13, 23].

The first example of producing an activated carbon by a combination of thermal and chemical processes was constituted by Bussy In 1822 who demonstrated that the decolorizing abilities of carbons depended on:

- the source material,
- the thermal processing,
- and the particle size of the finished product.

At the beginning of the twentieth century, Raphael von Ostrejko who patented between 1900 and 1903 made a revolution by exploring two distinct methods for the production of nanoporous carbon materials (activated carbon materials) from the activation of charcoal. This scientific breakthrough caused an improving the performance of these carbon materials in many applications by formation a high porosity in carbon materials skeleton.

Because of these discoveries, the first factory for the production of activated carbon materials has been built in Ratibor and was became the oldest factory for activated carbon production in the world [28, 30].

The first application of activated carbon was in World War I, when it was used in manufacturing soldiers masks for protection against hazardous gases and vapors [25].

The production and search for new activated carbons have been boosted decade after a decade due to their fundamental role in various technological applications which are related to, namely, restricted environmental regulations, recovery of valuable chemical compounds, and catalyst support. Nowadays, the driving forces for the research in nanoporous carbons are related to the properties of the most recent carbon materials: fullerenes, carbon nanotubes, and graphene. However, the excellent properties of these novel carbon forms also fostered the interest in the traditional porous carbons and, in recent years, a considerable number of studies searching for new synthetic approaches have been published. The main objective is the preparation of highly porous materials with controlled porosity, and often also with tuned surface chemistry, to present enhanced behavior as, for example, electrode materials for supercapacitors [26, 28, 31, 32].

4. Properties of nanoporous carbons

Carbon is one of the most abundant elements on the Earth and plays a critical role in the bio- and ecosystems. Carbon has the unique capability of forming a variety of interesting materials exhibiting extraordinarily different physical and chemical properties [20, 22, 27]. Fullerene [33], carbonnanotubes [33], graphite [15, 34], and diamond [35] are examples. To improve performance, nanoporous structures have been introduced into carbon because nanopores can give a large surface area.

Porous materials have various properties than bulk materials have [27].

- pores are classified according to shapes: cylindrical, spherical, and slit types [4, 29].

- pores are classified according to their accessibility to surroundings into (**Figure 1**):

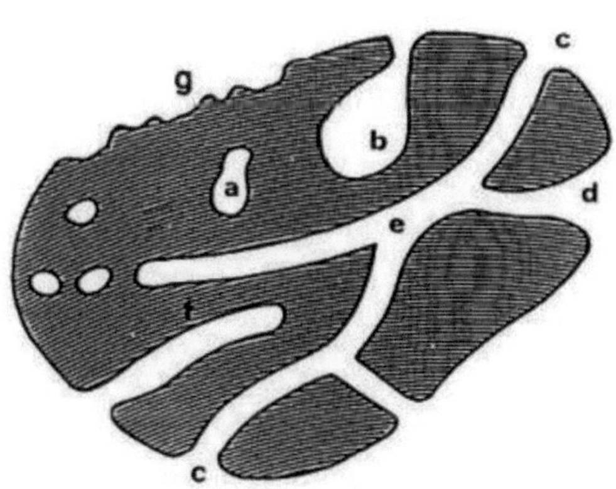

Figure 1.
Types of pores according to their accessibility to surroundings [29].

- open pores (b,c,d,e,f): pores that interact with the outer surface of the material and they are divided into pores open at one end (b,f) and pores that are open at two ends (e), and these type of pores are used in adsorption, catalysis and sensing processes [29].

- closed pores(a): pores which have no connection to the outer surface (isolated from the surroundings), they result from breakdown the parts close to the external shell of pores due to insufficient heating of porous materials, and these type of pores are used in Lightweight structural applications or thermal insulation [29, 36, 37].

IUPAC (International Union of Pure and Applied Chemistry) proposed the classification of pores according to their size:

- Micropores: pores that have a diameter < 2 nm, they are divided into pores their width less than 0.7 called ultramicropores or narrow micropores and pores in a range of 0.7 and 2 nm called supermicropores or micropores [4, 28].

- mesopores: pores that have a diameter between 2 to 50 nm.

- macropores: pores which have diameter > 50 nm [4, 19, 29, 36–38].

Nanoporous materials are materials with pore size in the range of 1-100 nm [15, 21, 29].

Nanoporous materials have unique features such as high specific surface area, shape-selective effects, fluid permeability, large porosity, and ordered uniform pore configuration. Therefore these materials can be used for many purposes such as separation, sensing, and catalysis applications [19, 39].

	Polymeric	Carbon	Glass	Alumino-silicate	Oxides	Metal
Surface area/ porosity	Low/larger than 0.6	High/in range 0.3–0.6	Low/in range 0.3–0.6	High/in range 0.3–0.7	Medium/in range 0.3–0.6	Low/in range 0.1–0.7
Pore size	Meso - Macro	Micro – meso	Meso-macro	Micro-meso	Micro-meso	Meso-macro
Strength	Medium	Low	Strong	Weak	Weak-medium	Strong
Permeability	low-medium	low-medium	High	Weak	low-medium	High
Thermal stability	Low	High	Good	Medium-high	Medium-High	High
Chemical stability	low-medium	High	High	High	Very High	High
Costs	Low	High	High	Low–medium	Medium	medium
Life	Short	Long	Long	Medium-long	Long	Long

Table 2.
Properties of nanoporous materials [39].

Various nanoporous materials with different properties such as surface area, porosity, pore size, thermal stability, etc. [9] are discussed in (**Table 2**).

The classification of pores discussed above is limited by the data of nitrogen adsorption–desorption at 77 k and that depends on: each pore size has a different mechanism of pore filling determined by isotherm profile [28].

1. Micropore filling: is represented as a primary physisorption divided into two categories:

    - ultramicropores (narrow micropores) filling occurs at low relative pressures ($P/P0 < 0.01$) and is controlled completely by the enhanced fluid–solid Adsorption interactions (enhancement of the adsorbent-adsorbate interaction). This process is called (primary micropore filling).

    - supermicropores (wider micropores) filling occurs at a higher relative pressure (P/P0 in range of 0.01–0.15) and is controlled by cooperative fluid–solid interactions and fluid–fluid interactions.

2. mesopores filling: occurs through the pore condensation (all adsorbate molecules are in contact with the surface of the adsorbent) followed by the occurrence of multilayer adsorption occurrence and end with capillary condensation due to the sorption process in mesopores relay on fluid–solid interaction and attractive fluid–fluid interactions.

3. The macropores are very large science act as open space, therefore, do not allow the capillary condensation [28, 40].

4. In nanoporous carbons, the porosity results from the spaces between crystallites graphite randomly cross-linked that form the carbon skeleton structure, less ordered carbon materials consumption, and from reactive carbon atoms removal in the crystallite during activation process [28].

4.1 Surface chemistry of nanoporous carbons

The main component of the nanoporous carbon skeleton is carbon atoms, but the basic structure of these materials also contains hydrogen and oxygen and may also include groups containing nitrogen, sulfur, or phosphorus, depending on the precursor, preparation route, and post-synthesis functionalization. Owing to the presence of unsaturated carbon atoms that are extremely reactive, these heteroatoms are primarily found at the edges of the basal planes. Due to particular interactions with the adsorptive and also the solvent in the case of solution adsorption, the elemental composition, and type of surface groups of a nanoporous carbon affect its efficiency in both gaseous and liquid phase processes. Properties such as hydrophobicity/hydrophilicity or acid/base action are extremely dependent on the surface chemistry of these materials [28, 37, 41, 42].

According to acid/base character, due to the presence of both acid and basic sites on their surface, nanoporous carbons are considered amphoteric materials. Thus, the materials may present net acid, basic or neutral surfaces depending on the amount and the power of all the surface groups [19, 38, 42].

Many methods can be used to evaluate nanoporous carbons surface chemistry and the best way to achieve a good characterization is using the supplementary techniques and incorporation between of the data analysis such as:

- Boehm titrations and potentiometric titrations give qualitative and quantitative data on the nanoporous carbon's surface.
- diffuse reflectance infrared spectroscopy (DRIFT) and X-ray photoelectron spectroscopy (XPS) give only qualitative information about the surface of the nanoporous carbons.
- and although with less quantitative information temperature-programmed desorption (TPD) detects more oxygen groups than Boehm titration [28, 38, 42].

4.1.1 Acidic surfaces

The chemical nature of nanoporous carbons is determined by surface groups containing oxygen that are mostly located on the external surface or edges of the basal plane.

The amount of oxygen on the surface has a high effect on the nanoporous carbons's adsorption abilities as these groups constitute the majority of adsorption surface.

These groups can be classified according to chemical nature into three categories: acidic, basic, neutral.

Carboxylic, lactone, phenol, carbonyl, pyrone, chromene, quinone, and ether groups are examples of oxygen-containing functional groups on the nanoporous carbons surface see (**Figure 2**).

The responsible for surface acidity is Functional groups such as:

Carboxylic acid or carboxylic anhydride, lactone, and phenolic hydroxyl.

These Oxygen-containing functionalities are created by oxidation of carbon surface. The most commonly used activation methods to introduce oxygen-containing acidic groups are oxidation by gases and aqueous oxidants.

- Gas-phase treatment: Oxygen, air, carbon dioxide, and steam can be used in the gas phase treatment. In these processes two routes of oxidation are used:
    - o oxidation at low temperature can be used to form strong acidic groups (carboxylic).
    - o oxidation at high temperatures can be used to form a large number of weak acid groups (phenolic).

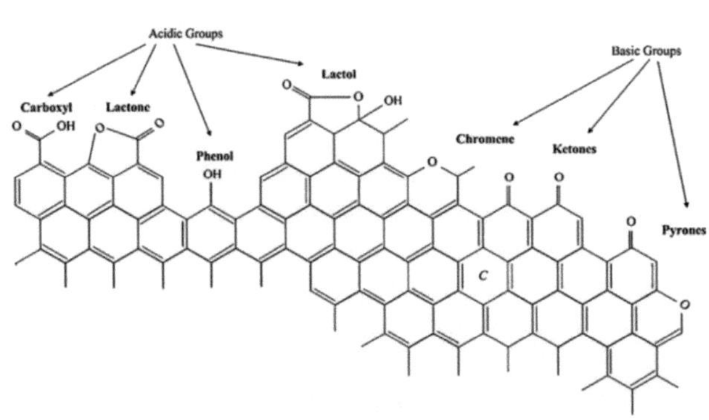

Figure 2.
Acidic and basic surface functional groups on a carbon basal plane [42].

- Liquid phase Treatment: Nitric acid or nitric and sulfuric acid mixture are very effective oxidizing agents due to the introduction of a significant number of oxygenated acidic functionalities onto the carbon surface that mainly includes carboxylic, lactone, and phenolic hydroxyl groups.

A greater quantity of oxygen groups in form of carboxylic and phenolic hydroxyl groups are produced in liquid phase oxidation at much lower temperatures compared to the gas phase oxidation [37, 42].

4.1.2 Basic surfaces

Basicity of activated carbon can be associated with:

i. resonating -electrons of carbon aromatic rings that attract protons,

ii. basic surface functionalities (e.g., nitrogen-containing groups) that are capable of binding with protons.

Chromene, ketone, and pyrone are oxygen-containing surface groups that respond to the nanoporous carbons' basicity (**Figure 2**).

The basic character of activated carbons, however, arises primarily from electrons of delocalized graphene-layer. It was proved that these electrons could act as Lewis bases.

The contribution of basal planes to carbon fundamentality has been studied by some researchers. Leon y Leon et al. studied the surface basicity of two carbon series and showed that solution protons can be adsorbed from oxygen-free carbon sites.

These sites are found on the basal plane of carbon crystallites in -electron-rich areas. Fundamental sites are therefore the Lewis type associated with the carbon structure itself [42].

Nitrogen-containing functionalities can be introduced through:

- either reaction with nitrogen-containing reagents (such as NH_3, nitric acid, and amines).
- or activation with nitrogen-containing precursors.

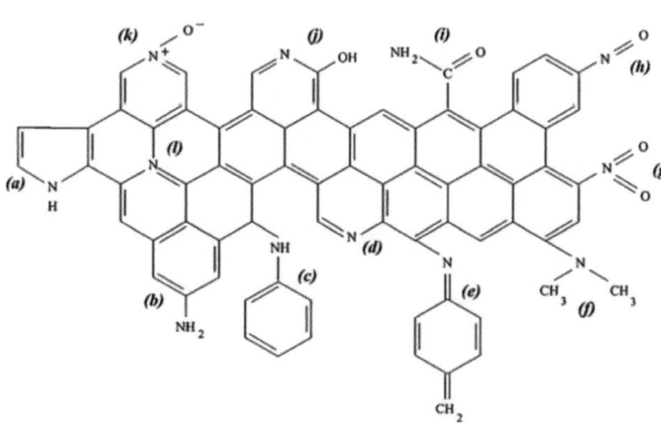

Figure 3.
Types of nitrogen surface functional groups: (a) pyrrole, (b) primary amine, (c) secondary amine, (d) pyridine, (e) imine, (f) tertiary amine, (g) nitro, (h) nitroso, (i) amide, (j) pyridone, (k) pyridine-N-oxide, (l) quaternary nitrogen [42].

Possible structures of the nitrogen functionalities include the following: amide group, imide group, lactame group, pyrrolic group, and pyridinic group; which are shown in (**Figure 3**). Nitrogen functionalities generally provide basic property, which can enhance the interaction between carbon surface and acid molecules such as, dipole–dipole, H-bonding, covalent bonding, and so on [37, 41, 42].

4.2 Nanoporous carbons analysis

Nanoporous carbons are used in various applications such as separation, catalysis and energy storage, and so on [19]. The properties of these materials depend on the application used. Therefore, characterization of these materials is very necessary to determine the properties of materials before use in experimental applications. Surface area, pore size, and porosity are important properties in the fields of catalysis, separation, batteries, gas and energy storage, and others.

As selectivity, diffusional rates and transport phenomena are important properties in catalyzed reactions, determining the pore structure in-depth is very necessary as it controls these properties.

Various techniques can be used for this purpose such as:

- gas adsorption analysis(physical adsorption)
- small-angle X-ray (SAXS)
- small-angle neutron scattering (SANS)
- Mercury intrusion porosimetry
- Nuclear magnetic resonance (NMR-based methods)
- scanning electron microscopy
- transmission electron microscopy
- thermoporometry,
- Brunauer, Emmett, and Teller (BET) technique

Each approach has a small applicability length scale for the study of pore size. The International Union of Pure and Applied Chemistry (IUPAC) gave a detailed overview of the various methods for characterizing pore size and their application range [43–45].

Gas adsorption is a common one among these, as it allows a wide variety of pore sizes to be examined, including the full range of micro-and mesopores. Moreover, as opposed to some of the methods described above, gas adsorption techniques are easy to use, are not harmful, and are not expensive [45].

4.2.1 Adsorption process

In general, adsorption is defined as the enrichment of molecules, atoms, or ions in the vicinity of an interface. In the case of gas/solid systems, adsorption takes place in the vicinity of the solid surface and outside the solid structure. The material in the adsorbed state is known as the (adsorbate), while the adsorptive is the same component in the fluid phase. The adsorption space is the space occupied by the adsorbate. Adsorption can be physical (physisorption) or chemical (chemisorption) [1, 5, 46].

4.2.1.1 Chemical adsorption: (chemisorption)

In chemisorption, the intermolecular forces involved lead to the formation of chemical bonds. When the molecules of the adsorptive penetrate the surface layer and enter the structure of the bulk solid, the term absorption is used. It is sometimes difficult or impossible to distinguish between adsorption and absorption: it is then convenient to use the wider term sorption which embraces both phenomena, and to use the derived terms sorbent, sorbate, and sorptive [1, 46].

4.2.1.2 Physisorption: (physical adsorption)

Is widely used for the surface and textural characterization of nanoporous materials (e.g. for textbooks and reviews see: Sanghi, Canevesi, Celzard, Thommes [40], and Fierro. 2020 [4]; Thommes 2015 [46]; Thommes 2014 [40]) [40]. Physisorption is a general phenomenon and occurs whenever an adsorbable gas (the adsorptive) is brought into contact with the surface of a solid (the adsorbent). The forces involved are the van der Waals forces. Physisorption in porous materials is governed by the interplay between the strength of fluid–wall and fluid–fluid interactions as well as the effects of confined pore space on the state and thermodynamic stability of fluids in narrow pores [6, 7, 40].

There has been considerable progress over the last two decades in understanding sorption phenomena in small pores, which in turn has contributed to substantial progress in physical adsorption.

The development and application of microscopic methods, such as functional density theory (DFT) of inhomogeneous fluids (e.g., nonlocal density functional theory, NLDFT) or computer simulation methods such as Monte Carlo (MC) and molecular dynamic (MD) simulations, is closely correlated with this advancement [29]. Among many porous materials, nanoporous carbons (NPC), with interpenetrated and regular nanopore systems, have recently triggered enormous research activities because of their fascinating chemical and physical properties, such as high specific surface area, tunable pore structure, catalytic activity, high thermal and chemical stability, intrinsic high electrical conductivity, low density, and wide availability. Therefore, they have been implemented in hydrogen storage, pollutant adsorption, energy storage,(i.e., batteries, supercapacitor), catalysts, energy conversion, and electrochemical devices [19, 27, 36, 37].

5. Nanoporous carbon materials synthesis

Various natural biomass such as cassava peel waste, chicken eggshell, seed shell, rubberwood sawdust, wood, peanut kernel, lignocellulose (biomass) materials, corn cob, Kraft lignin, scrap tires, textile waste, rice husk, palm shell, and sugar have been employed as precursors for the production of NPC. These sources are generally rich in carbon giving amorphous phases, and the plant wastes containing the cellulose are familiar to form graphitic nanostructures with high-temperature treatments [47].

Conventional porous carbon materials, such as activated carbon, have routinely been prepared by pyrolysis followed by the activation process of the organic precursors, such as coal, plant, wood, or polymers, at specific high temperatures [19].

5.1 Activation

Carbonaceous materials are activated to create porosity, controlled morphology, and functional groups on the surface. The pyrolysis process is generally carried out before undergoing the activation process as the former process generates organic residues, which may block the porous channels of the final carbon materials. Physical and chemical activation are two preferred choices for the fabrication of nanoporous carbon materials from carbon-rich precursors, including waste materials [47, 48] (**Table 3**).

5.2 Physical or thermal activation

Physical activation is usually carried out In two consecutive heating stages: Carbonization of the raw material under the inert atmosphere (usually nitrogen) to devolatilize the raw material, accompanied by activation consisting of partial gasification of the char acquired with oxidizing agents (i.e. steam, carbon dioxide or a combination of both) leading to the creation of a porous network.

While carbonization normally occurs at temperatures between 400 and 600 °C, temperatures ranging from 800 to 1000 °C are needed for gasification.

It is also possible to skip the carbonization stage, depending on the raw material, and proceed directly to thermal activation [47, 48].

- CO_2 is considered the preferred choice for physical activation due to the ease of handling, control of various parameters, and slow reaction rate. Instead of

Type of activation	Activating agent	Suitable precursors	Type of Porosity	The general trend of experimental conditions on pore size distribution (PSD)
Physical	Co_2	Coals and, to less extend, hard lignocellulosic materials	Micro only	• The high degree of activation contributes to a high size of micropores with a similar pore sizes distribution
	Steam	Coals and, to less extend, hard lignocellulosic materials	Micro and meso	• High activation degree leads to Widening of pore sizes distribution and obtaining on Micro and mesopore networks
Chemical	$ZnCl_2$	High volatile and oxygen content materials (Lignocellulosic materials)	Micro and meso	• Uniform micropore size distribution which broadening with the increase of the Zn/precursor ratio to the micro/mesopore boundary
	$H_3\ PO_4$	High volatile and oxygen content materials (Lignocellulosic materials)	Micro and meso	• Pore Size Distribution primarily in the border of micro/mesopore boundary and based on the temperature of heat treatment (<450 °C).
	KOH	Low volatile and high carbon content materials (coals of the High-rank)	Micro only	• The KOH/precursor ratio has a greater effect on the capacity of adsorption and PSD than the activation temperature. • Increasing the KOH/precursor ratio widens pores from narrow to large micropores and to small mesopores in a lesser extent; also hinders the morphology of granules (particle disintegration leads to powders).

Table 3.
Appropriate precursors, kinetic of activation, and type of porosity are typically obtained for the most common activating agents [28].

diffusional regulation which is quicker but contributes to external particle burning and, ultimately, to poor production of porosity. CO2 activation must occur in conditions that ensure chemical control (slow activation rate-days).

The reactions of steam and carbon dioxide with carbon are endothermic, thus:

To sustain the necessary high temperature, thermal activation requires an external energy supply [28, 31].

- Oxygen (or air) is not widely used as an oxidizing agent because its carbon reaction is highly exothermic and rapid, and instead of particle consumption, it is difficult to monitor and ensure porosity growth.

  Oxygen activation is scarcely used because of this and the safety concerns associated with temperature regulation.

  However, low amounts of oxygen (or air) may be added to steam or carbon dioxide during thermal activation to help sustain high temperatures by responding to the gases emitted during activation (i.e. CO and H2).

  This strategy has the benefit of decreasing CO and H2 pressure, both inhibiting activation gases and increasing the triggering agent's partial pressure [28] (**Table 3**).

5.3 Chemical activation

Chemical activation normally needs just one heating step: the raw material is combined with an activating agent (e.g. ZnCl2, H3 PO4, KOH) and further treated at temperatures between 400 and 900 °C under a controlled atmosphere, depending on the activating agent selected. The activating agent helps to remove the residual water moieties from the raw materials by acting as a dehydrating agent and also assists as an oxidant. Both the processes affect the decomposition of precursors and rearrangement of the resulting carbon atoms into an aromatic framework (**Table 3**).

Chemical activation offers an additional advantage of introducing functional groups such as -COOH, -NH, or -OH on the surface of the porous carbon. However, the crystallinity of the sample after the chemical activation is reduced due to the continuous dehydration and the oxidation with the activating agent, which creates a lot of defect sites along the carbon walls of the final product [28, 47, 48].

The mechanism of pore formation depends, in this process, on the chemical agent:

- Zinc chloride facilitates the elimination of water molecules from the raw material's lignocellulosic structures.

- Chemically, phosphoric acid merges with them.

- The selective removal of carbon atoms happens in none of these systems.

- The method is more complicated with potassium hydroxide as the structure is disintegrated and the metallic potassium is intercalated into the "graphic" laminar structure, particles are broken down and granular activated carbon synthesis is prohibited. At the same time, due to reaction with CO2 and H2 O, resulting from the redox reaction of carbon with potassium compounds, there is also some gasification of carbon atoms. The lignocellulosic precursor loses volume by contraction during carbonization (i.e. heat treatment under inert

	Chemical activation	Physical activation
Advantages	• One step • Energy saving • Higher yields • Shorter activation time (hours) • High surface area • High pore size.	• Utilizing mainly CO_2 and steam activation • Has a high yield • Has a high bulk density
Dis advantages	• Harsh reaction condition • Low recovery efficiencies • Time-consuming (washing to remove chemical residue. • Cost of required chemicals and processes Residue. • Some activating agents are harmful ($ZnCl_2$ and H_3PO_4). • Ununiformed pore distribution. • Difficult control of pore size.	• Relative low surface area and pore volume due to the lower degree of carbon etching. • Two consecutive heating steps

Table 4.
Advantages and dis advantages of activation processes [28, 48].

atmosphere), but when chemical activation is applied, the activating reagent is incorporated into the particles which inhibit the anticipated contraction, i.e. the activating agent will act as a template for microporosity formation. [28, 31, 32, 48, 49].

Chemical activation has advantages over the physical process discussed in (**Table 4**).

5.4 Other nanoporous carbons synthesis methods

The chemical activation process using KOH, K2CO3, K2O, ZnCl2, KHCO3, H_3PO4, etc., and their reaction are very useful to make nanopores, however, harsh reaction condition, cost of required chemicals and processes, residue, ununiformed pore distribution, and difficult control of pore size should be considered for upscaling the production and commercialization. Compared with chemical activation, physical activations utilizing mainly CO2 and steam activation usually has a high yield and bulk density but suffers from a relatively low surface area and pore volume due to the lower degree of carbon etching. Therefore, many researchers have studied the efficiency of other methods to fabricate nanoporous carbon materials.

Thus, hard- and soft-templating approaches have been successfully introduced for the preparation of NPC with well-defined pore structures and narrow pore size distributions. In this Chapter, the hard- and soft-templating synthesis are introduced as potential approaches for the preparation of NPC materials with a special emphasis on the progress and developments in the methodology.

Hard synthesis of templates requires the use of pre-synthesized organic or inorganic templates, while the soft synthesis of templates depends on the creation of nanostructures through self-assembling organic molecules [19, 23].

5.4.1 Templating method

Historically, Knox and co-workers, who demonstrated the synthesis of graphitic porous carbons for liquid chromatography separation by impregnation of spherical porous silica gel particles with phenolic resin and subsequent carbonization and silica removal, first reported the templating process in 1986. This technique has gained considerable attention since then and different types of template carbons are synthesized. The resulting carbon synthesized by the templating method has a relatively narrow PSD and regulated architecture called a templated carbon.

The templated carbonization method permits one to control the carbon structure in terms of various aspects, such as pore structure, specific surface area, microscopic morphology, and graphitizability, which makes this method very attractive [48].

Porous materials are fabricated in several different ways. The Hard Template Method and Soft Template Method are the two most common methods to make porous materials [39, 50, 51].

5.4.1.1 Hard template method

The most common hard template synthetic route for mesoporous carbon materials was first reported by Knox et al. using a spherical solid gel as the template. Highly ordered NPC with oriented mesoporous structures can be obtained using the hard template method.

The hard template method includes the following steps: (a) synthesis of a suitable porous template; (b) introduction of a suitable carbon precursor into the template pores using the method of wet impregnation, chemical vapor deposition, or a combination of both methods; (c) polymerization and carbonization of the carbon precursor; and (d) removal of the inorganic template. Following these steps, porous carbon with a specific pore structure is formed [19, 48, 52].

Angelina Sterczyńska, Małgorzata Śliwińska-Bartkowiak, Małgorzata Zienkiewicz-Strzałka, Anna Deryło-Marczewska Synthesized Nanoporous Carbon (also called ordered mesoporous carbon material [OMC]) with a 4.6 nm pore size, and ordered silica porous matrix, SBA-15, with a 5.3 nm pore size [54].

Also, Dandan Guo, Jin Qian, Ranran Xin, Zhen Zhang, Wei Jiang, Gengshen Hu, Maohong Fan prepared Mesoporous carbons enriched with nitrogen by hard template method for supercapacitors. Where CCl_4 and ethylenediamine (EDA)

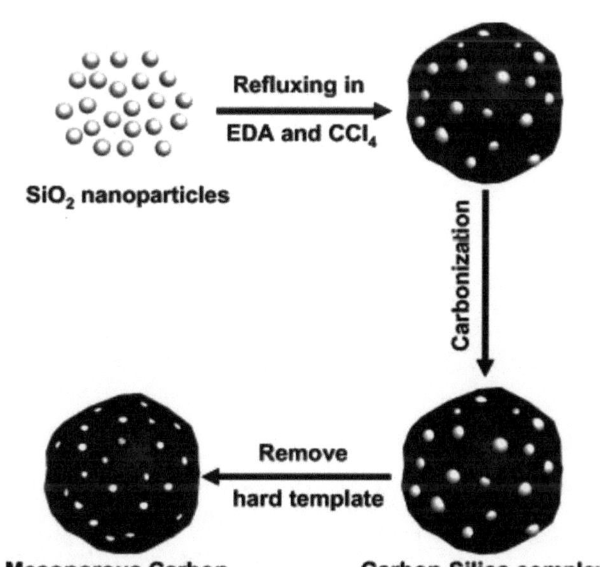

Figure 4.
Preparation of mesoporous carbon using silica porous matrix [53].

represent precursors whereas silica act as a hard template [53] (see **Figure 4**). While Wei Liu, Hong Yuan, and Yihu Ke Prepared ordered mesoporous carbon-based on soybean oil by using the hard template method where a hard template is represented by ordered mesoporous SiO_2 molecular sieves (SBA-15) [55].

However, when extracting from the template, the sacrificing of the solid template and mesoporous NPC structures limits the usefulness of hard template synthesis. The use of soft template synthesis will overcome these constraints [51, 54, 56].

5.4.1.2 The soft template method

The soft template is a kind of surfactant, which has a strong interaction with the carbon source, and mesoporous carbons with different structures can be obtained through the soft template method. This method possesses good controllability and operability; as a result, it has very good application prospects. The mechanisms of the soft template method include a liquid-crystal template mechanism, a synergistic assembly mechanism, a "rod micellar" mechanism, and so on; these mechanisms have been widely recognized [48, 52].

Amphiphilic molecules, such as surfactants and block copolymers, have been extensively used as soft-templates in the synthesis of ordered mesoporous materials. The discovery of ordered mesoporous carbon materials appeared to have a great impact in this field because of the fascinating features of their unique physical and chemical properties which can surmount the shortcomings in various technological applications. Preparing these ordered mesoporous carbons can be difficult to achieve by a simple self-assembly method for many reasons, although recent reports have demonstrated that polymeric micelles can serve as templates for mesoporous carbons. The key requirements for a successful synthesis using the soft-templating method are (i) the ability of the precursor species, such as the copolymers and the carbon source, to self-assemble into nanostructured polymer composite, (ii) the presence of at least one performing species, and one carbon source, (iii) the stability of the pore-forming species which can endure the temperature required for thermally decomposing the

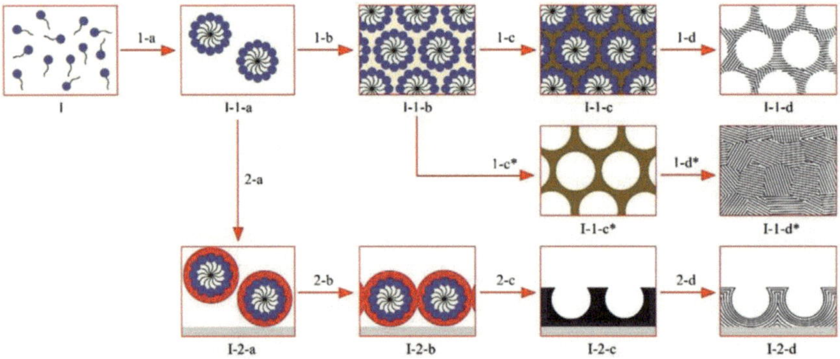

Figure 5.
Soft-templating synthesis of carbon nitride and graphene materials. Route 1: (1-a) self-assembly of surfactant or block copolymer molecules (I) into micelles (I-1-a), (1-b) addition of a carbon nitride precursor and formation of micelle-precursor mesostructures (I-1-b), (1-c) initial condensation/ polymerization of the precursor, (1-d) further condensation and template elimination creating a nanoporous carbon nitride material (I-1-d), (1-c) initial condensation of the precursor conducted at a temperature higher than the decomposition point of the soft template, (1-d*) further condensation but causing structural collapse. Route 2: (2-a) addition of a carbon precursor and production of individual micelle-carbon precursor units, (2-b) close-packing of these units on a substrate (side view, depicted in gray) forming a monolayer (1–2-b), (2-c) polymerization of the carbon precursor followed by carbonization, (2-d) graphitization giving nanoporous graphene sheets. Black lines represent the 2D building units, namely carbon nitride or graphene layers [19].*

carbon source during carbonization process, and finally (iv) the ability of the carbon source to form cross-linked polymers that can retain their nanostructures during the thermal decomposition. The synthesis principles of these self-assembled nanostructured mesoporous carbons open the way for the development of new strategies for materials in the future. Researchers have reported that only a few materials meet the requirements for the successful synthesis of ordered mesoporous carbons using a soft-templating approach [23, 50, 56]. Some of the research activities related to the soft-templating synthesis of polymeric structures are summarized [19] (see **Figure 5**).

6. Activated carbon as the essential phenol removal adsorbent

6.1 The activated carbon precursors

There is a wide range of raw materials that can be successfully used as a precursor for the preparation of activated carbon. Almost interesting precursors have been obtained from any carbonaceous materials such as agricultural waste, wood, petroleum coke, and industrial biomass. An important aspect in the preparation of activated carbon is the use of different parts of plants including the pulp, stems, shells, peels, flowers, fruits, seeds, stones, peels, and leaves. All these precursors can be carbonized and then activated under desired conditions to yield activated carbon [25]. The selection of the precursors is based mainly upon the following several factors:

- High carbon content and low amount of ash.
- Availability and inexpensive.
- Low content of inorganic matter.
- Nonhazardous for nature.

Raw Material	Carbon (mass%)	Volatiles (mass%)	Density ($cm^3 g^{-1}$)	Ash (mass%)	The texture of activated carbon
Soft wood	40–45	55–60	0.4–0.5	0.3–1.1	Soft, large pore volume
Hard wood	40–42	55–60	0.55–0.8	0.3–1.2	Soft, large pore volume
Lignin	35–40	58–60	0.3–0.4	—	Soft, large pore volume
Nutshells	40–45	55–60	1.40	—	Hard, large micropore volume
Lignite	55–70	25–40	1.0–1.35	5–6	Hard, small pore volume
Soft coal	65–80	20–30	1.25–1.5	2–12	Medium hard, medium pore volume
Petroleum coke	70–85	15–20	1.35	0.5–0.7	Medium hard, medium pore volume
Semi-hard coal	70–75	10–15	1.45	5–15	Hard, large pore volume
Hard coal	85–95	5–15	1.5–1.8	2–15	Hard, large pore volume

Table 5.
Properties of some raw materials and the properties of activated carbon generated [24].

The characteristic of activated carbons such as physicochemical properties that responsible for carbon adsorption properties and other possible applications depend on selected carbon precursors in addition to the preparation method. Lately, it is longer than lignocellulosic resources and waste biomass is the most used precursors for the production of the activated carbon (**Table 5**). Summary of the properties of some raw materials and the properties of activated carbon generated [24, 57, 58].

The usage of lignocellulosic biomass in the generation of activated carbon has many features as it is renewable and most abundant in nature, inexpensive, and helps to dispose of its negative impact effect on the environment. Numerous reviews have been devoted to inexpensive precursors of activated carbon in recent years [58, 59].

6.2 Generation of porosity and surface chemistry (activated carbon)

All activated carbon is generally characterized by a porous structure with their high surface area, usually have few amounts of chemically bounded heteroatom oxygen, hydrogen, sulfur, and nitrogen. Beside may contain around 20% by weight of a mineral substance called ash content [24, 49]. It is known that the surface of activated carbon has a high heterogeneous phenomenon. AC surface heterogeneity comes from two various sources called geometrical and chemical sources. Geometric heterogeneity results from differences in the size and shape of pores, cracks, pits, and steps. Chemical heterogeneity is associated with different functional groups, especially the oxygen groups that are most often found at the edges of turbine crystals, as well as with various surface impurities. The heterogeneity of AC (Geometric, Chemical) surfaces affect unique adsorption properties. The chemical properties and structure of activated carbon and its structure can be changed depending on the type and nature of presence and number of oxygen functional groups on its surface [24, 25, 36, 59].

It's possible to produce activated carbon from all carbonaceous materials which its preparations involve two major steps: carbonization of the precursors followed by activation method as shown in (**Figure 6**). Carbonization means the conversions of raw materials at elevated temperatures into a highly stable carbon structure with an elementary and partially -developed pore structure. During this step, water and volatile substances are removed leaving the char behind. Followed by activation of char by physical or chemical activation to produce highly porous activated carbon. The generated activated carbon characterize by having a porous structure, high

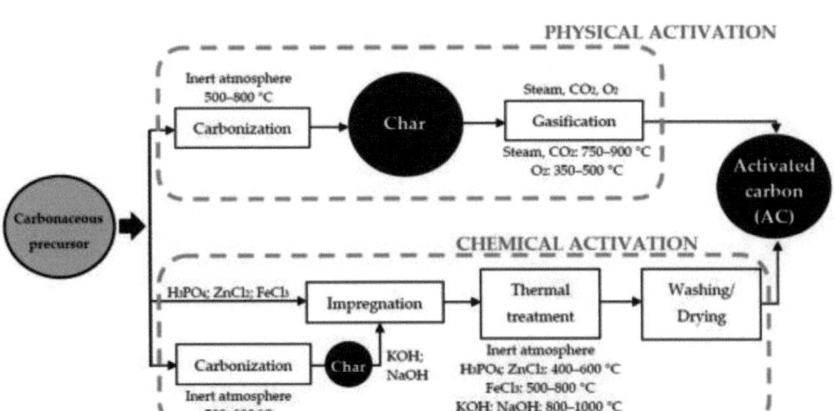

Figure 6.
Synthesis of activated carbon [49].

surface area, and highly reactive surface functionality. Physical activation involves the carbonization of the precursors in the presence of inert gas in the range between 500 to 900 °C followed by gasification of the resulting char with carbon dioxide. Steam, air, or mixtures of both can be also used as activating agents. Excessive temperatures lead to reduced carbon content, collapse its pore structures and increase ash generation. in physical activation, proper temperature and time are so important to achieve adequate pore development and the creation of functional groups. In chemical activation, the raw material is directly impregnated with an activating agent such as KOH, NaOH, H_3PO_4, H_2SO_4, HNO_3, $ZnCL_2$, and $FeCL_3$, and the impregnated product is pyrolyzed at high temperature for a certain time and then product washed to remove the activating agent. The activating agent can contribute in the oxidation and gasification the carbon precursors to improve porosity and transform surface functional group. The pores generated from activation are usually identified as microspores and mesoporous [24, 28, 48]. Chemical activation has advantages among physical included in (a)-the low temperature of activation. (b) Well developed in the porous structures.

Different pore sizes (micro, meso, and macropores) are obtained depending on the nature and type of precursors, activating agent, and reaction conditions such as time and temperature. The properties of raw material such as its type and size, the type of activating agent, the ratio of mixing raw material with the activating agent, the conditions of heating in the furnace, will have a significant effect on the characteristics of the final product including surface area and pore size [24, 48].

However, first and foremost, the adsorption features of activated carbon are dictated by its chemical composition. The existence of hydrogen and oxygen groups on the surface of the activated carbon directly affects the adsorption performance. Original AC precursors, the activation process, or post-treatment after the preparation process can be the source of these surface groups. The oxygen groups are mainly formed on the surface of activated carbon following activation by air exposure or by relevant post-treatment [25, 57, 60, 61].

Carbon is more likely to chemisorb oxygen than any other species. Chemisorbed oxygen present on the surface of AC to form carbon and oxygen functional groups can be acidic, neutral, or basic. The formation of oxygen groups on the carbon surface is generated from the reaction with the activating agent used such as (H_2So_4, HNO_3, H_2O_2) and other oxidizing gases like CO_2 and O_3. Among the factors affecting the nature of the surface group of oxygen, the temperature may be taken into account, as surface acidity is formed that includes carboxyl functional groups, carboxylic anhydrides, lactones, and phenol hydroxyls upon exposure to low temperature while basic surface that generated from delocalized π-electrons on a carbon basal plane like pyrone, quinone, and carbonyl generated at high tempertature [24].

To define the number of groups of oxygenated surfaces, The Boehm titration method is used. The basic functional groups are the most preferred than the acidic functional group for the adsorption of phenolic compounds. And some experimental methods like temperature-programmed desorption (TPD), infrared spectroscopy, acid–base titration, X-ray photoelectron spectroscopy (XPS), can be used to characterize surface-oxygen groups [24, 57].

6.3 Role of surface heterogeneity on adsorption of phenol

Although activated carbon has been investigated for a long time as an effective adsorbent of organic pollutants, the exact structure of the functional groups and the mechanism of phenolic compounds adsorption is not well understood yet.

Much information should be considered before applied adsorption of phenol such as:

1. The effects of surface functionalities on adsorption phenols exhibit a complex significance than the porosity effect.

2. Activated carbon have an amphoteric character in an aqueous solution (possesses both, acidic and basic surface functional groups) and has a positive or negative charge on its surface depending on the solution of PH. The type of nature of activated carbon surface affects the adsorption of organic electrolytes such as phenol so it is so important to determine the charge of the surface of activated carbon as well as the extent of ionization of the phenol before the application of phenol adsorption [41].

3. The chemical characteristics on activated carbon surface determine from functional groups and π- delocalized electrons of fused aromatic structures. The affinity of the activated carbon toward adsorbate can be determined from the content of surface functional groups and pHpzc. The pHpzc indicates zero net surface charge of the adsorbents that implies their electronic surface charges. The surface is positively charged at $pH < pH_{PZC}$ in which water gives protons more than the hydroxyl group and when $pH > pH_{pzc}$ the surface has a negative charge. It is commonly assumed, that for $pH < pK_a$ adsorption of non-ionized organics does not depend on the surface charge of AC. However, for $pH > pK_a$ the phenolic compound is dissociated, and adsorption of its ionic form depends on the surface charge.

4. In the case of phenolic adsorption, the basic surface of activated carbon is so preferred to achieve high performance of adsorption of phenol.

The adsorption capacity of phenol on activated carbon depends on some factors such as:

a. The solubility of phenols in water.

b. The degree of activation.

c. The hydrophobicity of substituted phenols [24, 25, 42].

And thus adsorption capacity increase with increasing specific surface area and porosity while it decreases by the solubility of phenolic compounds in water and increases the hydrophobicity of phenolic substituted. For example, phenolic compounds that having low solubility in water like p-cresol and p-nitrophenol are adsorbed on activated carbon than other phenols. On the other hand, chlorophenol, nitrophenol, cresol adsorbed greater on activated carbon than phenol and aminophenol due to their hydrophobic group. The adsorption of phenolic compounds onto the ACs mainly contribute to three types of interactions namely, (i) π-π dispersion interaction, (ii) the electron-donor–acceptor complex formation, and (iii) the hydrogen-bonding formation. The mechanism of adsorption of phenolic compounds toward activated carbon occurs through the formation of electron donor-acceptor complex between the aromatic ring of phenol and basic sites on the surface of activated carbon (basic surface oxygen complex and/or πelectron-rich sites on the basal planes). Therefore, the relative affinity between the carbon surface's basic characteristic and aromatic phenolic ring increases. Electron withdrawing of phenolic rings tends to form electron donor-acceptor complex between these ring and basic sites on the surface of activated carbon [24, 25, 42]. In the case of oxidation, the surface of activated carbon with a strong oxidizing agent leads to the formation of the acidic surface with a large quantity of carboxyl and phenolic groups with a small amount of carbonyl and chromene lead to inhibition of phenol adsorption. During the adsorption of phenol on activated carbon, these regions act as a donor and the aromatic rings of phenol as acceptors. Phenol adsorption onto the activated carbon is controlled by dispersive force between π electrons. The interaction of π-π dispersion occurs between basal planes of activated carbon and the phenol aromatic ring [24, 42]. The change in PH solution affects phenol adsorption. The adsorbed amount of PH decrease at low and high PH values. At low PH value, protons were added to compete with the adsorbate for the carbonyl sites leading to a reduction of adsorption of phenol at this value. Besides the surface chemistry of activated carbon, the pore structure also affects the adsorption process. The porosity of activated carbon has been considered an important factor in the adsorption processes of phenolic compounds from aqueous solutions. The adsorption capacity of small molecules such as phenol to the inner surface of carbon correlates with the content of micropores and BET surface area, while for mesoporous ACs, substituent group in the phenol and nature of the carbon controlled the phenol adsorption as well [24, 37, 38, 41].

7. Conclusion

Nanoporous carbon materials have an attractive rate performance in many applications of recent technology such as pollutant adsorption. In this chapter, the properties of nanoporous carbon and its various preparation methods are presented. Also, our choice of the preparation method, reaction conditions, and the precursor materials affect the properties of the resulting nanoporous structure. The adsorption of phenolic compounds from polluted water is one of the most common uses of nanoporous carbon, especially activated carbon in water treatment. Numerous factors are known to have an important influence on phenolic adsorption like the type of carbon structure, functional groups present on the surface, oxygen availability on its surface, pH value of the aqueous media, etc. Furthermore, there are several scientific papers reviewed - aspects most relevant to indicating today's trends and potential insights in elucidating the adsorption mechanisms of phenolic compounds on activated carbon.

Appendices and nomenclature

Nanoporous carbon materials: (NPC).

Author details

Mahmoud Fathy Mubarak[1*], Alshimaa Maher Ahmed[2] and Sahar saad Gabr[1]

1 Petroleum Application Department, Egyptian Petroleum Research Institute, Cairo, Egypt

2 Department of Chemistry, Faculty of Science, Helwan University, Cairo, Egypt

*Address all correspondence to: fathy8753@gmail.com

References

[1] Singh, S., et al. (2020). Chapter 10 - Low-cost adsorbents for removal of inorganic impurities from wastewater. Inorganic Pollutants in Water. P. Devi, P. Singh and S. K. Kansal, Elsevier: 173–203.

[2] Eduardo Enrique Pérez Ramírez, Miguel de la Luz Asunción, Veronica Saucedo Rivalcoba, Ana Laura Martínez Hernández and Carlos Velasco Santos (March 15th 2017). Removal of Phenolic Compounds from Water by Adsorption and Photocatalysis, Phenolic Compounds - Natural Sources, Importance and Applications, Marcos Soto-Hernandez, Mariana Palma-Tenango and Maria del Rosario Garcia-Mateos, IntechOpen, DOI: 10.5772/66895. Available from: https://www.intechopen.com/books/phenolic-compounds-natural-sources-importance-and-applications/removal-of-phenolic-compounds-from-water-by-adsorption-and-photocatalysis.

[3] William W. Anku, Messai A. Mamo and Penny P. Govender (March 15th 2017). Phenolic Compounds in Water: Sources, Reactivity, Toxicity and Treatment Methods, Phenolic Compounds - Natural Sources, Importance and Applications, Marcos Soto-Hernandez, Mariana Palma-Tenango and Maria del Rosario Garcia-Mateos, IntechOpen, DOI: 10.5772/66927.

[4] Sdanghi G, Canevesi RLS, Celzard A, Thommes M, Fierro V. Characterization of Carbon Materials for Hydrogen Storage and Compression. C. 2020; 6(3):46. https://doi.org/10.3390/c6030046

[5] Wang, J., et al. (2020). "Degradation of phenolic compounds by dielectric barrier plasma: Process optimization and influence of phenol substituents." Chemical Engineering Journal **385**: 123732.

[6] Tri, N. L. M., et al. (2020). "Removal of phenolic compounds from wastewaters by using synthesized Fe-nano zeolite." Journal of Water Process Engineering 33: 101070.

[7] Asencios, Y. J. O., et al. (2020). "Removal of phenol in seawater by heterogeneous photocatalysis using activated carbon materials modified with TiO2." Catalysis Today.

[8] Villegas, L. G. C., et al. (2016). "A Short Review of Techniques for Phenol Removal from Wastewater." Current Pollution Reports 2(3): 157–167.

[9] Nakagawa, K., et al. (2004). "Adsorption of phenol and reactive dye from aqueous solution on activated carbons derived from solid wastes." Water Research 38(7): 1791–1798.

[10] Turco A, Monteduro AG, Mazzotta E, Maruccio G, Malitesta C. An Innovative Porous Nanocomposite Material for the Removal of Phenolic Compounds from Aqueous Solutions. Nanomaterials. 2018; 8(5):334. https://doi.org/10.3390/nano8050334

[11] Chen, M., et al. (2018). "Methane adsorption behavior on shale matrix at in-situ pressure and temperature conditions: Measurement and modeling." Fuel 228: 39–49.

[12] Duan, J., et al. (2021). "Simultaneous adsorption of uranium (VI) and 2-chlorophenol by activated carbon fiber supported/modified titanate nanotubes (TNTs/ACF): Effectiveness and synergistic effects." Chemical Engineering Journal 406: 126752.

[13] Jiang, L., et al. (2020). "Role of adsorption and oxidation in porous carbon aerogel/persulfate system for non-radical degradation of organic

contaminant." Chemosphere 241: 125066.

[14] Kong, L. and H. Adidharma (2018). "Adsorption of simple square-well fluids in slit nanopores: Modeling based on Generalized van der Waals partition function and Monte Carlo simulation." Chemical Engineering Science 177: 323–332.

[15] Tofighy, M. A., et al. (2020). Chapter 6 - Development of advanced nanocomposite membranes by carbon-based nanomaterials (CNTs and GO). Nanocomposite Membranes for Water and Gas Separation. M. Sadrzadeh and T. Mohammadi, Elsevier: 145–162.

[16] Singh, S., et al. (2020). Chapter 10 - Low-cost adsorbents for removal of inorganic impurities from wastewater. Inorganic Pollutants in Water. P. Devi, P. Singh and S. K. Kansal, Elsevier: 173–203.

[17] Gusain, R., et al. (2019). "Recent advances in carbon nanomaterial-based adsorbents for water purification." Coordination Chemistry Reviews 405.

[18] Gopalakrishnan, K. (2015). Nanocarbons for Advanced Energy Storage, Volume 1, DO - 10.1002/9783527680054.ch8.

[19] Malgras, V., et al. (2019). "Fabrication of Nanoporous Carbon Materials with Hard- and Soft-Templating Approaches: A Review." J Nanosci Nanotechnol 19(7): 3673–3685.

[20] Yu, L., et al. (2020). "Solvent-free synthesis of N-doped nanoporous carbon materials as durable high-performance pH-universal ORR catalysts." Journal of Colloid and Interface Science 575: 406–415.

[21] Van Riet, R., et al. (2020). "Novel opportunities for nanoporous carbons as energetic materials." Carbon 164: 129–132.

[22] Gomis-Berenguer, A., et al. (2016). "Sulfur-mediated photochemical energy harvesting in nanoporous carbons." Carbon 104: 253–259.

[23] Haynes, T., et al. (2020). "Preparation of mesoporous silica nanocapsules with a high specific surface area by hard and soft dual templating approach: Application to biomass valorization catalysis." Microporous and Mesoporous Materials 306: 110400.

[24] Dąbrowski, A., et al. (2005). "Adsorption of phenolic compounds by activated carbon—a critical review." Chemosphere 58(8): 1049–1070.

[25] Water and Wastewater Treatment: Historical Perspective of Activated Carbon Adsorption and its Integration with Biological Processes. Activated Carbon for Water and Wastewater Treatment: 1–11.

[26] Hassan, A. F. and H. Elhadidy (2017). "Production of activated carbons from waste carpets and its application in methylene blue adsorption: Kinetic and thermodynamic studies." Journal of Environmental Chemical Engineering 5(1): 955–963.

[27] Marpaung, Freddy, Synthesis of nanoporous carbon derived from hybrid metal organic frameworks for symmetric supercapacitor, Master of Philosophy thesis, Institute for Superconducting and Electronic Materials, University of Wollongong, 2019. https://ro.uow.edu.au/theses1/780

[28] Ana S. Mestre and Ana P. Carvalho (December 20th 2017). Nanoporous Carbon Synthesis: An Old Story with Exciting New Chapters, Porosity - Process, Technologies and Applications, Taher Hcine Ghrib, IntechOpen, DOI: 10.5772/intechopen.72476. Available from: https://www.intechopen.com/books/porosity-process-technologies-and-applications/nanoporous-carbon-

synthesis-an-old-story-with-exciting-new-chapters

[29] Zdravkov, B. D., et al. (2007). "Pore classification in the characterization of porous materials: A perspective." Central European Journal of Chemistry 5(4): 1158–1158. The original version of the article was published in Cent. Eur. J. Chem., Vol. 5(2), (2007), pp. 385–395. It can be also found online at: http://dx.doi.org/10.2478/s11532-007-0017-9.

[30] Kerimkulova, A., et al. (2017). "Synthesis and application of carbon adsorbents in chromatographic separation of biologically active complexes." Frontiers in Nanoscience and Nanotechnology 3.

[31] Luo, H., et al. (2018). "Self-terminated activation for high-yield production of N,P-codoped nanoporous carbon as an efficient metal-free electrocatalyst for Zn-air battery." Carbon 128: 97–105.

[32] Huang, J., et al. (2020). "In-situ synchronous carbonation and self-activation of biochar/geopolymer composite membrane: Enhanced catalyst for oxidative degradation of tetracycline in water." Chemical Engineering Journal 397: 125528.

[33] Sirivisoot, S. and T. J. Webster (2011). 5.523 - Carbon Nanotubes: Applications for In Situ Implant Sensors. Comprehensive Biomaterials. P. Ducheyne. Oxford, Elsevier: 303–315.

[34] Mortazavi, B., et al. (2019). "Nanoporous graphene: A 2D semiconductor with anisotropic mechanical, optical and thermal conduction properties." Carbon 147: 377–384.

[35] Yu, K.-l., et al. (2006). "Synthesis of NiO-embedded carbon nanotubes using corona discharge enhanced chemical vapor deposition." Diamond and Related Materials 15(9): 1217–1222.

[36] Noshadi, I., et al. (2016). "Porous carbonaceous solid acids derived from farm animal waste and their use in catalyzing biomass transformation." Applied Catalysis A: General 513: 19–29.

[37] Yoo, D., et al. (2019). "Porosity control of nanoporous CuO by polymer confinement effect." Scripta Materialia 162: 58–62.

[38] Lowry, E. and M. Piri (2018). "Effects of chemical and physical heterogeneity on confined phase behavior in nanopores." Microporous and Mesoporous Materials 263: 53–61.

[39] Toyin Jibowu(2016) A Review on Nanoporous Metals. Front Nanosci Nanotech,Volume 2(4): 165–168 doi: 10.15761/FNN.1000129.

[40] Thommes, M. and K. A. Cychosz (2014). "Physical adsorption characterization of nanoporous materials: progress and challenges." Adsorption 20(2): 233–250.

[41] Lowry, E. and M. Piri (2018). "Effects of chemical and physical heterogeneity on confined phase behavior in nanopores." Microporous and Mesoporous Materials 263: 53–61.

[42] Shafeeyan, M. S., et al. (2010). "A review on surface modification of activated carbon for carbon dioxide adsorption." Journal of Analytical and Applied Pyrolysis 89: 143–151.

[43] Xiaoguang Yang & Shaobin Guo (2020): Comparative analysis of shale pore size characterization methods, Petroleum Science and Technology, DOI: 10.1080/10916466.2020.1780258.

[44] Anovitz, L. M. and D. R. Cole (2015). "Characterization and Analysis of Porosity and Pore Structures." Reviews in Mineralogy and Geochemistry 80(1): 61–164.

[45] Cychosz, K. A., et al. (2017). "Recent advances in the textural

characterization of hierarchically structured nanoporous materials." Chemical Society Reviews 46(2): 389–414.

[46] Thommes, M., et al. (2015). “Physisorption of gases, with special reference to the evaluation of surface area and pore size distribution (IUPAC Technical Report)." Pure and Applied Chemistry 87(9–10): 1051–1069.

[47] Stalin Joseph, Gopalan Saianand, Mercy R. Benzigar, Kavitha Ramadass, Gurwinder Singh, Anantha-Iyengar Gopalan, Jae Hun Yang, Toshiyuki Mori, Ala'a H. Al-Muhtaseb, Jiabao Yi, Ajayan Vinu. Recent Advances in Functiona lized Nanoporous Carbons Derived f rom Waste Resources and Their Applica tions in Energy and Environment. Advanced Sustainable Systems. (2020) 2000169 10.1002/adsu.202000169

[48] Zhou, J., et al. (2019). CHAPTER 1 Carbon-based CO2 Adsorbents. Post-combustion Carbon Dioxide Capture Materials, The Royal Society of Chemistry: 1–75.

[49] Bedia, J., et al. (2020). “Review on Activated Carbons by Chemical Activation with FeCl3." C 6(2).

[50] Gras, R., et al. (2006). “Template synthesis of carbon nanotubes from porous alumina matrix on silicon." Microelectronic Engineering 83(11): 2432–2436.

[51] Hwang, S., et al. (2007). “Template-directed synthesis of highly ordered nanoporous graphitic carbon nitride through polymerization of cyanamide." Applied Surface Science 253(13): 5656–5659.

[52] Zhang, L. and M. Jaroniec (2020). “Strategies for development of nanoporous materials with 2D building units." Chemical Society Reviews 49 (16): 6039–6055.

[53] Guo, D., et al. (2019). “Facile synthesis of nitrogen-enriched nanoporous carbon materials for high-performance supercapacitors." Journal of Colloid and Interface Science 538: 199-208

[54] Au - Sterczyńska, A., et al. (2019). “Surface Properties of Synthesized Nanoporous Carbon and Silica Matrices." JoVE(145): e58395.

[55] Liu, W., et al. (2020). “Preparation and characterization of ordered mesoporous carbon based on soybean oil." Journal of Materials Science 55(15): 6525–6536.

[56] Fan, J., et al. (2007). “Hard-templating synthesis of a novel rod-like nanoporous calcium phosphate bioceramics and their capacity as antibiotic carriers." Materials Chemistry and Physics 103(2): 489–493.

[57] Freitas, J. V., et al. (2019). “Coconut shell activated carbon as an alternative adsorbent of inhibitors from lignocellulosic biomass pretreatment." Industrial Crops and Products 137: 16–23.

[58] Rodriguez-Reinoso, F. and J. Silvestre-Albero (2016). Activated Carbon and Adsorption. Reference Module in Materials Science and Materials Engineering, Elsevier.

[59] Li, B., et al. (2020). “Adsorptive removal and mechanism of monocyclic aromatics by activated carbons from water: Effects of structure and surface chemistry." Colloids and Surfaces A: Physicochemical and Engineering Aspects 605: 125346.

[60] Kim, Y. I., et al. (2019). “High-capacitance activated bio-carbons with controlled pore size distribution for sustainable energy storage." Journal of Power Sources 438: 226969.

[61] Lütke, S. F., et al. (2019). “Preparation of activated carbon from black wattle bark waste and its application for phenol adsorption." Journal of Environmental Chemical Engineering 7(5): 103396.

Chapter 9

Graphene Nanopores

Per A. Löthman

Abstract

Graphene is a two-dimensional, atomic thin, usually *impermeable* nanomaterial with astonishing electrical, magnetic and mechanical properties and can therefore at its own right be found in applications as sensors, energy storage or reinforcement in composite materials. By introducing *nanoscale pores* graphene alter and extend its properties beyond permeability. Graphene then resembles a nanoporous sensor, a nanoporous, atomic thin membrane which opens up for such varied applications such as water purification, industrial waste water treatment, mineral recovery, analytical chemistry separation, molecular size exclusion and supramolecular separations. Due to its nanoscopic size it can serve as nanofilters for ion separation even at ultralow nano- or picomolar concentrations. It is an obvious choice for DNA translocation, reading of the sequence of nucleotides in a DNA molecule, and other single molecular analyses as well for biomedical nanoscopic devices since dimensions of conventional membranes does not suffice in those applications. Even though graphene nanopores are known to be unstable against filling by carbon adatoms they can be stabilized by dangling bond bridging via impurity or foreign atoms resulting in a robust nanoporous material. Finally, graphene's already exceptional electronic properties, its charge carriers exhibit an unusual high mobility and ballistic transport even at 300 K, can be made even more favorable by the presence of nanopores; the semimetallic graphene turns into a semiconductor. In the pores, semiconductor bands with an energy gap of one electron volt coexist with localized states. This may enable applications such as nanoscopic transistors.

Keywords: graphene, carbon, DNA, bottom-up, translocation, sculpting

1. Introduction

Pores are *ubiquitous* in nature, engineering and the natural sciences. We recall such diverse examples such as porous light weight metals, aluminum foams or metallic hollow spheres structures [1–4] that can save energy by reduced gas consumption, or cell membranes with ion-channels which constitute highly functional nanoporous structures of the cell. They are responsible for maintaining the required pressure gradient, ion-flux and ultimately nutrition and life itself in both mono- and multicellular organisms. One of the simplest but highly ordered proteinaceous nanoporous membranes in nature are bacterial surface layers (s-layers), a spontaneously occurring protective layer on the surface of the bacterial cell. This regularly structured nanoporous membrane protects and regulates a minimum out- and influx of nutrients to the cell and can be used as templates to synthesize metallic nanoparticles in the lower nanorange (1,4 nm) or as a soft membrane in nanobiodevices [5]. Nanopores such as the protein hemolysin are found in cell membranes, acting as transport channels for ions or molecules in and out of cells [6–8]. The selection mechanism of these membranes can be based on size exclusion as well as exclusion based on double layer overlap and dielectric exclusion [9]. S-layers are together with the cell membrane one of numerous *soft-matter* nanoporous materials. While cell membrane ion-channels or s-layers serve several life sustaining, protective biological functions they are not durable or mechanical stable and therefore not suitable for engineering applications. Quite the opposite are *solid state* nanopores which presents obvious advantages over their soft-matter counterparts. They are highly stable, exhibit controllable dimensional parameters such as channel length and diameter. Their surface characteristics can be altered and enable integration into devices and arrays.

A solid state material which has gained considerable recognition the last decade is *graphene* considered as one of the strongest and thinnest materials known. Monolayer Graphene possesses astonishing characteristics: Its electron mobility is 100 times higher than silicon; it conducts heat twice as good as diamond; its electrical conductivity is 13 times better than copper and it absorbs only 2.3% of reflecting light i.e. it is transparent; it is impenetrable even to the extent that the smallest atom (helium) cannot pass through a defect-free monolayer graphene sheet; and its high surface area of 2630 m^2/g which means that with less than 3 grams you could fully cover an entire soccer field. It is a two-dimensional atomic thin allotrope of carbon consisting of a single layer of atoms arranged in a two-dimensional hexagonal honeycomb structure [10, 11]. The name reflects the fact that the graphite allotrope of carbon consists of stacked graphene layers [12]. They are bound to each other by weak van der Waal forces which makes graphene an integral part of the 3D material graphite from which it was first isolated. Graphene was however not expected to exist in the free state i.e. as a single monoatomic layer. Scientists had argued convincingly that monoatomic thin 2D materials like graphene would be too thermodynamically unstable to exist. Thermal fluctuations would be as large as the force binding the atoms together, causing the structure to fall apart [10, 11]. However, carbon bonds are in fact strong enough and small enough that thermal fluctuations are not enough to destabilize graphene even at room temperature. Free-standing monolayer graphene was isolated in 2004 by Novoselov and Geim [13, 14] and follow-up investigations revealed several novel exciting properties [15, 16]. Graphene was considered as the new material of the future and Novoselov and Geim were awarded the Nobel Prize in Physics for their discovery and characterization of graphene.

Graphene has excellent *mechanical and electrical* properties: an atomic thin monolayer graphene has an inplane direction independent Young's modulus of 1 TPa and strength of 100 GPa [17–21].

Graphene's superior electrical properties are due to the fact that its charge carriers are massless Dirac fermions [10–12] with high mobility and ballistic transport even at highest electric-field and affected to only minor degree by chemical doping. Its extraordinary high electrical conductivity and its capacity to carry large currents at room temperature [22] makes it indeed an exciting material. Carbon atoms have four electrons available to make chemical bonds. Graphene is however only one atom thick and every atom in the crystal is bound to only three others. Each atom thus has one free electron available for electronic conduction which means that graphene by far exceed the electrical properties of metals. Since each graphene 2D lattice provides as many charge carriers as metals are only able to supply from bulk 3D atomic architectures, even when metals tend to have some electrons delocalized and shared in a "sea of electrons" among all atoms within a piece of metal, which makes graphene an extraordinary material in electronics [23]. In electronics graphene may act as scaffold on which parts that can act as distinct components may self-assemble into an electronic circuit. This is due to the fact that various molecules can attach to the graphene surface due to its electronic structure and that chemical changes made to parts of the graphene sheet such that local electric properties can be fine-tuned and varied on the same surface along with additional properties such as permeability via nanopores (sculpting). Such nanometer-sized circuitry may one-day enable faster and smaller computational and electronic devices.

Moreover, the specific electrical properties of graphene in terms of conductivity are due to the fact that with one p_z electron per atom in the model the valence band is fully occupied, while the conduction band is vacant. The two bands touch at the zone corners (the K point in the Brillouin zone), where there is a zero density of states but *no band gap*. The graphene sheet thus displays a *semimetallic* (or *zero-gap-semiconductor*) character, although the same cannot be said of a graphene sheet rolled into a carbon nanotube, due to its curvature. By introducing nanopores in graphene one can open up an energy band gap in a graphene sheet as described below.

Conceptually graphene represents a new class of materials; inorganic, two-dimensional materials that are only one atom thin. Thereby graphene provides new incursions into low-dimensional physics which has always been a rich source for novel applications. Graphene does no longer requires any further proof of its importance in terms of fundamental physics, however, there is still room for extending, altering and improving graphene properties. As already mentioned *nanoscopic sculpting* such as nanolithography, manipulation by AFM [24] or an electron beam of a transmission electron microscope [25] is considered a promising venue to target properties of nanomaterials. In this way nanopores can be introduced into graphene. It has become an alternative route of materials development; instead of turning to a different class of graphene-based materials such as nanoribbons or nanocomposites, sculpting nanopores into graphene would further alter the already numerous and exceptional properties and extend the fields of applications. It would open up an energy band gap in a graphene sheet for the application as field effect transistors (FETs) [25–27]. Nanopores can turn *semimetallic* graphene into a *semiconductor* [28].

Nanoporous graphene exhibit a *periodic arrangement* with nanoscale diameters in the graphene membrane. Apart from Nano sculpting as mentioned above numerous methods, such as chemical etching [29], vapor deposition [30], and electron beam [31], have been developed to fabricate nanoporous materials and control pore dimensions. Moreover, it is expected that facile methods such as self-assembly of graphene are just as suitable for nanoporous graphene as it is for impermeable graphene or grapheneoxide [32–34].

The aim of this work is to illustrate how the properties and applications of the nanoscopic material graphene can be altered, improved and extended by introducing nanopores in the graphene layer.

2. Ion transport through graphene nanopores

Transport phenomena through ion exchange membranes have been investigated for several decades. When L. Michaelis first observed the effect of membrane charge on the ion transport through pores in the year 1926 [35, 36] there has been an continuous interest in this research field and the transport phenomena are now well understood for conventional dense ion exchange membranes. The current trend towards nanotechnology and miniaturization of devices, ion transport through solid state nanopores is gaining attention [37–41]. 2D materials such as graphene play an important role for applications in nanofluidic device, biosensing, and DNA translocation [42–45].

2D materials may have some limits in these applications because of the presence of intrinsic defects and low surface charge density. The ion selectivity may be influenced by the pore size distribution. These membranes exhibit low surface charge which limits rejection of ions. It is therefore important to optimize fabrication techniques combined with a thorough understanding of transport phenomena through a 2D interface. It is expected that transport under nano-confinement in 2D is expected to differ from highly charged ion exchange membranes.

A number of different physical transport processes occur in the pores of a membrane. The most relevant processes for ion transport and ion separation processes the most important are size exclusion, charge exclusion and dielectric exclusion [46]. Size exclusion occurs when the pore size of the membrane is comparable or smaller than the species to be retained.

Microfiltration (MF) membranes have relatively large pore sizes (0.1–10 μm), to separate smaller species from 1 to 100 nm (e.g. proteins, viruses), ultrafiltration (UF) membranes are used. Nanofiltration (NF) membranes (1–10 nm) are used for removal of salt, amino acid, and dye [47]. Dielectric exclusion is an ion rejection mechanism observed in NF membranes and typically dominates at <1 nm and effective up to about 2 nm pore size [46, 48]. This phenomena occurs at interfaces between media having different dielectric constants. The mutual interaction of ions at the surface and the induced bound electric charge at the interface leads to the dielectric exclusion. This also depends on pore geometry e.g. cylindrical pores have stronger exclusion compared to slit pores. Ion exchange membranes (IEM) are used for demineralization or deionization of water, energy conversion and energy storage in fuel cells, redox flow batteries [49, 50].

Other than these commercial membranes, nano-porous materials such as solid state nano-pores in synthetic membranes (SiN_x, SiO_2), *nano-porous graphene*, graphene oxide multi layers, metal organic frameworks (MOFs), zeolitic imidazolate frameworks (ZIFs), and hybrid membranes can act as *ion selective membranes* depending on the ion concentration.

Konatham *et al.* studied water transport through monolayer graphene nanopores via molecular dynamics (MD) simulations. The pore diameter war as small as 7.5–14.5 Å in the monolayer graphene [51]. In this case ion exclusion was achieved up to a 7.5 Å pore diameter of non-functionalized (uncharged) pores. Larger pores cannot block the ions. Dielectric exclusion may also in this case be an important mechanism of exclusion for pore sizes close to 7.5 Å.

The ion rejection mechanism in pores is influenced by functionalized pores. Functionalization with carboxyl groups show improved ion rejection due to a higher free energy barrier towards water and ions. Because of this ion screening effect, the free energy barrier decreases with increasing salt concentrations in the bulk. Cohen-Tanugi *et al.* showed via MD simulation that multilayer *graphene membranes* can desalinate water more effectively than monolayer graphene. The salt rejection mechanism as a function of pore diameter, layer spacing and applied pressure was investigated. The smaller nanopores (3 Å) reject salt entirely compared to larger pores (4.5 Å) and highly aligned pores with multiple layers can even combine high salt rejection with high water flux [52].

Graphene pores supported on track etched PCTE membranes and pores were enlarged by oxidative etching in acidic potassium permanganate solution was investigated. The number of pores were $1012/cm^2$ and the pore sizes were in the sub nanometer (0.40 ± 0.24 nm) range. For short oxidative etching times, the resulting membrane showed cation selective behavior because of steric exclusion and the negatively charged surface groups at the pores. A membrane potential around 4 mV was observed for 0.5 M KCl/0.1667 M KCl which is lower compared to the theoretical Nernst potential for this salt concentration ratio (28.1 mV). This indicates that the pore sizes may be larger than 0.4 nm as the selectivity is expected to be higher at 0.5 M KCl due to dielectric exclusion. The membrane potential decreased with increase in pore size obtained after longer oxidative etching times.

Oxygen (O_2) plasma etching is an additional fabrication method to create graphene nanopores; CVD graphene transferred onto a SiN substrate with a 5 μm hole and subjected to O_2 plasma etching. This generates nanometer sized holes in the graphene sheet which was confirmed by Raman spectroscopy. Nanopores fabricated via this method have shown very high *salt retention* at lower etching time due to the small pore size. The transport properties of ions through these nanoporous interfaces were investigated with or without externally applied electric fields [53–60].

Nanoporous graphene is *cation selective* [61, 62] and a non-linear current–voltage relationship has been detected [63]. A diode rectification effect has widely been observed especially in solid state conical nanopores [54, 56]. For graphene nanopores and other 2D materials (MoS_2, h-BN) the rectification effects has been observed in both intrinsic and artificially made pores [61, 64–68]. Nanoporous graphene supported on PET shows an ion rectification effect due to the presence of *conical* nanopores as a result of *asymmetric etching* [64–68]. Applying an external potential and gating the graphene the ion selectivity can even be *tuned* [69].

Up to date most investigations were limited to sub-nanometer sized pores in graphene where the ion rejection was mainly dominated by steric exclusion.

The ion-selectivity of graphene nanopores favors K^+ over Cl^- even up to a pore diameter of 20 nm. The selectivity calculated by the Goldman-Hodgkin-Katz (GHK) model was around 100, a value much higher than the selectivity observed by Jain *et al.* for pore sizes of sub-nm level (0.4 nm) [63]. The membrane potential, specially for biological membranes is typically calculated using the GHK voltage Equation [70–76]. This equation is applicable for multiple permeating monovalent species and it takes into account the permeability of each specie. Furthermore, the selectivity was dependent on the pH. The cation to anion selectivity decreases with decreasing pH which was attributed to the protonation of surface charged groups e.g. carboxyl groups at the graphene edge. Interestingly, the membrane could differentiate between monovalent and divalent cations by conducting monovalent cations 5 times faster than divalent cations. Ion selective transport through graphene with pores larger than a nanometer was experimentally shown by Rollings and van Deursen et al. [74, 75].

Molecular dynamics simulations by Cohen-Tanugi *et al.* show that nanoporous free-standing graphene membranes are able to reject NaCl ions while letting water flow at permeabilities several orders of magnitude higher than conventional reverse osmosis membranes. The performance was studied as a function of pore size, chemical functionalization, and applied pressure. The results indicate that the membrane's ability to prevent the salt passage but allowing for water flow depends critically on pore diameter. Also chemical functional groups bonded to the edges of graphene pores suggests that commonly occurring hydroxyl groups can roughly double the water flux thanks to their hydrophilic character. Nanoporous graphene may play an important role for water purification. The maximum diameter for salt permeability is around 5.5 Å, that is, Na + and Cl- ions will pass through the membrane beyond this diameter [77].

Other molecular dynamics studies by Suk *et al.* found that pure water can continue to flow across graphene nanopores with diameters below 1 nm, and calculations suggest that the chemical functionalization of graphene nanopores could be tuned to selectively reject certain solvated ions [78–81].

3. Single molecule analysis via graphene nanopores

Nanopores resemble a class of a biosensor, allowing for highly sensitive detection of biomolecules including nucleic acids and proteins at single-molecule resolution [76, 82, 83]. Nanopore sensors have emerged as powerful devices for probing biomolecules and offer a novel platform for single molecule analysis and characterization. In particular, they have attracted significant attention as tools for high-throughput, robust, and low-error DNA sequencing. Especially graphene and other two dimensional (2D) materials are being investigated with respect to their integration into nanoscaled devices that may in the future sequence genomes. The successful implementation of solid-state nanopores in emerging third-generation DNA sequencing applications is contingent upon developing methods for scalable fabrication, high-accuracy output, and integration with low-noise electronic architectures.

In a nanopore ion currents and forces can be monitored as molecules pass through. This makes it possible to investigate a wide range of phenomena involving DNA, RNA and proteins. The solid-state nanopore increasingly proves to be a surprisingly versatile new single-molecule tool in biophysics and nanofluidics.

The high sensitivity of the nanopore comes from the characteristic structure of the nanopore: a nanometer scale pore as large as the size of the molecule of interest [84, 85]. It allows for detection of biomolecules even at a sub-nanomolar concentration level and discrimination of minute differences in molecular structure between different nucleotides [86–88]. The high sensitivity originates from the electric potential applied across the nanopore membrane which generates a highly concentrated electric field near the nanopore. Charged molecules pass through the narrow pore one molecule at a time [87].

The methods in use for biomolecule translocation detection include resistive pulse sensing [89–92], tunneling current detection [42, 93, 94] and optical sensing, [95–99]. In the resistive pulse technique (or Coulter-counter method), the nanopore acts as the only channel across the membrane for both ions and biomolecules, and partial blocking of the nanopore by a biomolecule is directly reflected in a perturbation in the measured ionic current. More precisely, a membrane containing a single nanopore is sandwiched between two reservoirs of electrolytic solution, such as aqueous potassium chloride. Ions are driven through the pore, as illustrated in **Figure 1**, by applying an electric potential difference across the membrane, resulting in an ionic current that can be measured. When the electrolyte contains larger charged molecules, such as DNA or proteins, these are also driven through the pore, causing a transient dip in the ionic current where each current dip represents a passage, or translocation, of a biomolecule, with the magnitude and duration of the pulse being indicative of the molecule's radius and length, respectively.

Moreover, slowing down the DNA translocation speed has been a major issue for nanopore sensing. The DNA translocation speed is a few orders of magnitude faster in the solid-state nanopore than in a biological nanopore for unknown reasons [95].

To slow down translocation speed in the solid-state nanopore, various engineering strategies have been enivsaged: dragging the molecule by strengthened interaction with the modified or chemically decorated pore surface. Here materials including aluminum oxide (Al_2O_3)–graphene etc. integrated with the nanopores slows down the DNA translocation by an enhanced Coulombic, specific, or hydrophobic nonspecific interaction between the nanopore surface and DNA [101, 102].

Various research groups investigated the influence of the number of nanoporous graphene layers on the DNA translocation; 1 or 2 layers thick [103], 1–8 layers thick [101] and 3–15 layers thick [102]. Initial DNA detection experiments were carried out, an important step towards DNA sequencing. In each case the nanopores could

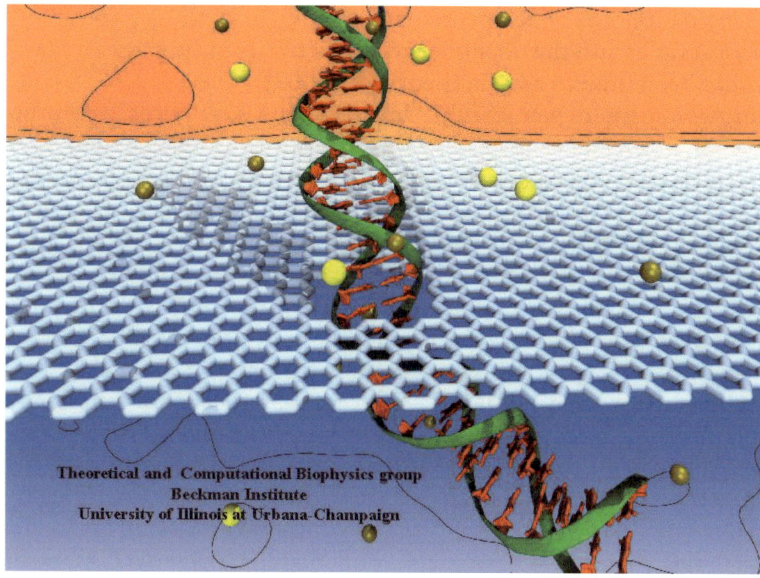

Figure 1.
A DNA molecule translocating through a graphene nanopore [100].

detect double stranded DNA molecules with lengths from 400 to 48,000 base pairs. Even membranes with significant variability in the baseline current levels were found to be viable for DNA detection. Interestingly the nanopores could differentiate between DNA that passed through the pore in an extended form and that which passed through in a folded form. Even though these developments are impressive, the central goal remains unsolved: is singlebase resolution with a graphene nanopore feasible? Also in this case all the different nanoporous graphene membranes did show that the translocation events are too fast to be resolved by the existing detection electronics.

4. Mechanical and electrical properties of nanoporous graphene

Nanoporous graphene is unique in that it exhibits both *electronic functionality* as a *tunable* semiconductor and *mechanical functionality* as a *tunable* molecular filter membrane. These properties combined in a single atomically-thin, mechanically robust platform makes nanoporous graphene a promising candidate for electronically active nanodevice applications [104–106].

As is noted, a better understanding of the structure–property relation would be of direct relevance to the structure design and function optimization in a variety of technological applications. The present study aims at the mechanical properties of nanoporous graphene membranes. Many attempts have been made to exploit the basic properties of nanoporous graphene membranes for functional applications. In this context Cohen-Tanugi *et al.* [105] used molecular dynamics simulations and continuum fracture mechanics in order to study the mechanical resilience of nanoporous graphene as a reverse osmosis membrane. The mechanical properties such as strength depend on the nanopore architecture in the nanoporous graphene materials and the nanopore diameter of the substrate .

An energy band can be opened by introducing nanopores in a graphene sheet for example for application as field effect transistors (FETs) [101, 102, 106]. Semimetallic graphene, the normal state of graphene, can turn into a semiconductor by introducing nanopores . The opening and tuning of a bandgap in nanoporous graphene membranes and the dependence of electronic properties on the structural parameters has been investigated in theory [102, 103].

The energy bandgap of semiconducting graphene nanopores, a chirality dependent scaling rules have been suggested. On the basis of extensive tight binding studies and simple geometric arguments, *Lee et al.* report that Pedersen scaling governs not only the energy bandgap but also the effective mass of the Bloch electron of the semiconducting graphene nanopores regardless of its chirality or the crystallography of pores when the nanopore areal fraction is low [102].

To open a tunable bandgap in graphene, which is required for semiconductor materials, has been desirable for novel applications of graphene. One strategy of constructing periodic nanopores in graphene to form graphene antidot lattices (GALs) has been extensively studied The electronic structure of graphene antidot lattices with zigzag hole edges was studied with first-principles calculations. It was revealed that half of the possible GAL patterns were unintentionally missed in the usual construction models used in earlier studies. With the complete models, the bandgap of the GALs was sensitive to the width W of the wall between the neighboring holes. A nonzero bandgap was opened in hexagonal GALs with even W, while the bandgap remained closed in those with odd W. Similar alternating gap opening/closing with W was also demonstrated in rhombohedral GALs. Moreover, analytical solutions of single-walled GALs were derived based on a tight-binding model to determine the location of the Dirac points and the energy dispersion, which confirmed the unique effect in GALs [103].

Hu *et al.* [101] investigated the mechanical behavior and fracture mechanism of nanoporous graphene NPG for porosities up to 80% and marked the transition of mechanical behavior at a critical porosity of ~15%. Carpenter *et al.* [102] analyzed the dependence of elastic properties on the architecture of graphene nanopore arrays (the pore arrangement, pore morphology, material density, and pore edge passivation), and further established the scaling law between modulus and relative density. Moreover, Liu *et al.* [106] carried out MD simulations to study the mechanical properties of nanoporous graphene with the pore size ranging from 0.4 nm to 1.3 nm, and for the first time revealed the relationships between mechanical properties (Young's modulus and fracture strength) and porosity. These investigations have a shed light upon deformation behavior and mechanical properties of nanoporous graphene.

5. Fabrication of graphene nanopores

Graphene seems to be an ideal material to create nanopores because it is mechanically and chemically robust even when being atomically thin. A defect free graphene layer is completely impermeable. For this reason, pore creation is necessary to investigate transport mechanisms through graphene nanopores as well as altered mechanical and electronic properties. Pore creation in this two dimensional material is challenging as it is difficult to handle this monolayer graphene without creating additional defects and cracks.

The first attempts to fabricate nanoporous graphene was based on the *top-down approach* in which the structures were created via electronbeam lithography (nano sculpting) or directly etched from graphene, patterned by using self-assembled etch masks as described in the upper part of **Figure 2**. The top-down method do however not lead to small nanopores that can open band gaps of roughly 1 eV (comparable to that in silicon, a conventional semiconductor material). The diameter should be less than 2 nm to meet this requirement [27, 108] for implementation in devices. Since this is beyond the structural resolution of top-down approaches successful attempts

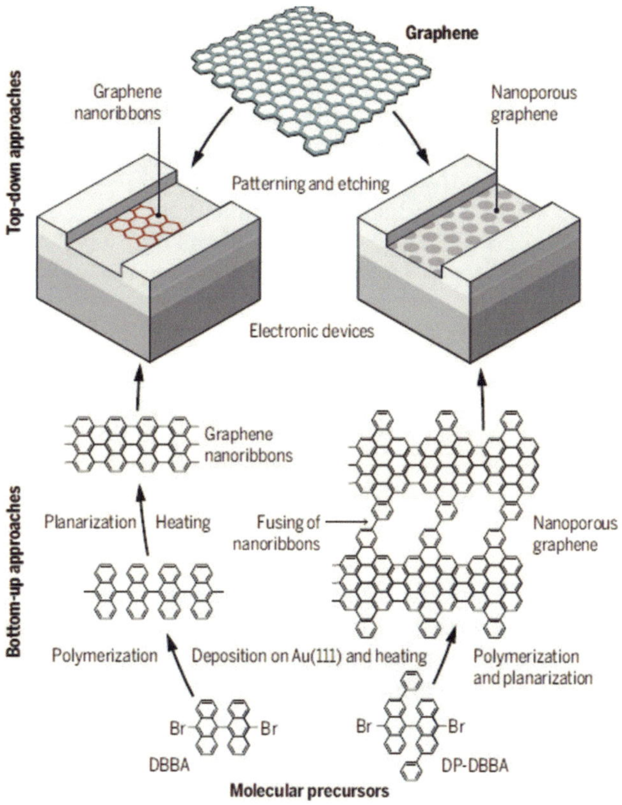

Figure 2.
Nanopores can be introduced in graphene by top-down or bottom-up strategies. Top-down approaches involve patterning and etching of graphene sheets. The resulting pore structure are often too large for device applications that require a band gap. Bottom-up assembly using molecular precursors can overcome this limitation as shown in the lower part [28, 107].

to create graphene nanopores of the desired size was rather achieved via *bottom-up* approaches (lower part of **Figure 2**). Starting from molecular building blocks, such as DBBA (10,109-dibromo-9,99-bianthracene) and other halogenated molecular precursors [28, 109] very narrow graphene nanoribbons could be made with atomic precision when sublimed onto a single crystal Au(111) substrate. The molecules form linear polymer chains in ultra-high vacuum (UHV) at about 200°C and annealing at 400°C where the chains planarize and fuse. Other halogenated polycyclic aromatic hydrocarbon precursors lead to a large variety of structurally different graphene nanoribbons via similar on surface reactions [102] and of the nanoribbons are not entirely straight i.e. a non- uniform width, prior to fusion, they produce graphene nanostructures with nanoscopic holes after an extra annealing at 450°C which is the case when diphenyl-substituted DBBA (DP-DBBA) is sublimed on Au(111) substrate in UHV. Several spectroscopic studies have shown that the resulting nanoporous graphene has a highly anisotropic electronic structure with a band gap of about 1 eV [110].

Additional bottom-up strategies use polyphenylene units through surface-assisted coupling of halogenated molecular building blocks [108, 109] as well as chevron-shaped graphene nanoribbons fused to form nanoscale graphene nanopores [110]. Other bottom-up methods such as chemical etching [18], vapor deposition [25], and electron beam [26], have been developed to fabricate NPG materials and control the characteristic size. Moreover, it is even possible to realize a *single nanopore* in a graphene sheet as Rollings *et al.* have shown via fabricating a single nanopore supported on SiNx by an electrical pulse method [74].

Graphene nanopores are the material of choice for applications in nanodevices and in this context it should be mentioned that characterization where graphene are implemented is still a challenge. The short length scales (mostly around <50 nm) as well as the necessity of accurate alignment relative to the device structure, and high contact resistances, and in studies of electrical properties of nanoribbons the yield of working devices is often rather low. On the other hand nanoporous graphene could form larger electrically conducting domains, from which devices for electrical property measurements could be produced at yield as high as ~75% yield [110, 111].

6. Conclusion

It is obvious that for improvement and alteration of graphene properties introducing nanopores into the graphene sheets can open up novel fields of application for graphene. Nanoporous graphene design is of equal importance in terms of improving long-range order of nanopores, avoiding defects and establish accurate dimensional control and increasing yield. Here the bottom-up approach seem to have advantages over the top-down approach. Hopefully the potential of nanoporous graphene will stimulate chemists to develop new molecular precursors for nanoporous graphenes with various combinations of structural parameters (size, geometry and arrangement of pores). To induce periodic, atomically-precise nanopores and to tailor precise dimensions and electronic properties and to fabricate nanoporous graphene with complete atomic precision is a future goal. However the present highly anisotropic structure of nanoporous graphenes may also be of interest for spectroscopic studies. Graphene nanostructures with nanoscopic pores may be of interest for applications such as separation, sensing, and potentially even DNA sequencing. Since nanopores in graphene open up the band gap and makes the material semiconducting promising applications in mostly FETs but also water filtration, supercapacitors, biological analysis, DNA translocation and molecular sieving among others can be envisaged [112, 113]. It is because of the extended plethora of properties of nanoporous graphene that it is considered to be the next leap forward in carbon based nanomaterials research.

Author details

Per A. Löthman
Foviatech GmbH, Hamburg, Germany

*Address all correspondence to: per.loethman@foviatech.com

References

[1] Waag, U., Schneider, L., Löthman, P.A., Stephani, G., *Metallic Hollow Spheres - Materials for the future*, Metall Powder Report, 55, January 2000

[2] Illerhaus, B., Jasuiniene, E., Goebbels, Löthman, P.A., *Investigation and image processing of cellular metals with highly resolving 3D micro-tomography (μCT)*, Proceedings of SPIE, Developments in X-Ray Tomography III, February 2002

[3] Löthman, P.A., Andersen, O., Gründer, J., Landgraf, G., *Mechanical properties of metallic hollow sphere structures and Characterisation of their Cell wall properties*, Materialwissenschaft u. Werkstofftechnik, June 2002

[4] Löthman, P.A., Waag, U., Schneider, L., Stephani, *Metallic Hollow Spheres and Hollow Sphere Structures for Automotive Applications*, Int. Conf. on Mat. Eng., New Materials Techn. and Biomaterials, Liberec, Czech Republic, June 2001

[5] Hüttl, R., Ullrich, F., Wolf, G., Kirchner, A., Löthman P.A., Katzschner, B., Pompe, W., Mertig, M., *Catalytic Carbon Monoxide Oxidation Using Bio-Templated Platinum Clusters*, Catalysis Letters, 132 (3), October 2009

[6] C. Cao and Y. T. Long, *Biological Nanopores: Confined Spaces for Electrochemical Single-Molecule Analysis*, Accounts of Chemical Research 51, 331 (2018).

[7] G. Di Muccio, A. E. Rossini, D. Di Marino, G. Zollo, and M. Chinappi, *Insights into Protein Sequencing With An Hemolysin Nanopore by Atomistic Simulations*, Scientific Reports 9, 6440 (2019).

[8] T. Diederichs, G. Pugh, A. Dorey, Y. Xing, J. R. Burns, Q. Hung Nguyen, M. Tornow, R. Tampé, and S. Howorka, *Synthetic Protein-Conductive Membrane Nanopores Built With DNA*, Nature Communications 10, 5018 (2019).

[9] Cha, J., Kyoung, W., *Molecular Dynamics Simulation of the Effects of Affinity of Functional Groups and Particle-size on the Behavior of A Graphene Sheet in Nanofluid*, Computational Materials Science 139, 202 (2017).

[10] Geim, A.K., & Novoselov, K.S. (2007) *The rise of graphene*. Nat. Mater. 6:183-191

[11] Zhang, Y., Tan, J. W., Stormer, H. L. & Kim, P., *Experimental observation of the quantum Hall effect and Berry's phase in graphene*. Nature 438, 201-204 (2005)

[12] Peres, N. M. R.; Ribeiro, R. M. (2009). *Focus on Graphene*. New Journal of Physics. 11 (9)

[13] Novoselov, K. S. *et al.*, *Electric field effect in atomically thin carbon films*. Science 306, 666-669 (2004)

[14] Novoselov, K. S., *Two-dimensional atomic crystals*. Proc. Natl Acad. Sci. USA, 102, 10451-10453 (2005).

[15] Novoselov, K. S., *et al.*, *Two-dimensional gas of massless Dirac fermions in graphene*. Nature, 438, 197-200 (2005).

[16] Schedin, F., Geim, A.K., Morozoy, S.V., Hill, E.W., Blake, P., Katsnelson, M.I., Novoselov, K.S., *Detection of individual gas molecules adsorbed on graphene*. Nature Materials 6, 652-655, (2007)

[17] Bunch, J.S., *et al.*, *Electromechanical resonators from graphene sheets*. Science, 2007. 315(5811): p. 490-493.

[18] Stankovich, S., *et al.*, *Graphene-based composite materials*. Nature, 2006. 442(7100): p. 282-286.

[19] Zhou, J., Huang, R., *Internal lattice relaxation of single-layer graphene under in-plane deformation.* Journal of the Mechanics and Physics of Solids, 2008. 56(4): p. 1609-1623.

[20] Mu, W., *Study of axial strain-induced torsion of single-wall carbon nanotubes using the 2D continuum anharmonic anisotropic elastic model*. New Journal of Physics, 2009. 11(11): p. 113049.

[21] Memarian, F., Fereidoon, A., Darvish Ganji, M. *Graphene Young's modulus: Molecular mechanics and DFT treatments,* Superlattices and Microstructures, Vol. 85, September 2015, p. 348-356

[22] Novoselov, K.S., *Graphene: The magic of flat carbon*. ECS Transactions 19(5): 3-7. 2009

[23] Geim, A.K. (2009) *Graphene: Status and prospects*. Science 324:1530-1534.

[24] Giesbers, A.J.M., Zeitler, U., Neubeck, S., Freutag, F., Novoselov, K.S., Maan, J.S., *Nanolithography and manipulation of graphene using an atomic force microscope*, Solid State Communications, 147, (9-10): 366-369, July 2008

[25] Kosynkin, D.V., Higginbotham, A.L., Sinitskii, A., Lomeda, J.R., Dimiev, A., Price, B.K., *et al.*, *Longitudinal unzipping of carbon nanotubes to form graphene nanoribbons*, Nature 458 (2009) 872e876.

[26] Jiao, L., Zhang, L., Wang, X., Diankov, G., Dai, H., *Narrow graphene nanoribbons from carbon nanotubes*, Nature 458 (2009)

[27] Pedersen, T.G., Flindt, C., Pedersen, J., Mortensen, N.A., Jauho, A.-P., Pedersen, K., *Graphene Antidot Lattices: Designed Defects and Spin Qubits,* PRL 100, 136804, 2008

[28] Moreno, C., Vilas-Varela, M., Kretz, B., Garcia-Lekue, A., Costache, M.V., Paradinas, M., Panighel, Ceballos, G., Valenzuela, S.O., Peña, D., Mugarza, A., *Bottom-up synthesis of multifunctional nanoporous graphene,* Science 360, 199-203, 2018

[29] Han, T.H., Huang, Y.K., Tan, A.T.L., David, V.P., Huang, J., *Steam etched porous graphene oxide network for chemical sensing*, J. Am. Chem. Soc. 133 (2011)

[30] Safron, N.S., Kim, M., Gopalan, P., Arnold, M.S., *Barrier-guided growth of micro- and nano-structured graphene*, Adv. Mater. 24 (2012)

[31] M.D. Fischbein, M. Drndic, *Electron beam nanosculpting of suspended graphene sheets*, Appl. Phys. Lett. 93 (2008) 113107.

[32] Löthman, P.A., *Graphene and Self-assembly,* Global Webinar on Materials Science and Engineering, July 09-11, 2021, Insights and Innovations in Materials Science & Engineering: Progressing to the Future

[33] Löthman, P.A., Hageman, T.A.G., Elwenspoek, M.C., Krijnen, G.J.M., Mastrangeli, M., Manz, A., Abelmann, L., *A Thermodynamic Description of Turbulence as a Source of Stochastic Kinetic Energy for 3D Self-Assembly,* Adv. Mater. Interfaces 2019, 1900963

[34] Löthman P.A., *Macroscopic Magnetic Self-assembly*, PhD Thesis, University of Twente, April 2018

[35] Sata, T., *Ion Exchange Membranes: Preparation, Characterization, Modification and Application*, The Royal Society of Chemistry, Cambridge, U.K., (2004)

[36] Michaelis L., Perlzweig, W.A., *Studies on Permeability of Membranes: I. Introduction and The Diffusion of Ions Across The Dried Collodion Membrane*, Journal of General Physiology 10, 575 (1927).

[37] R. M. M. Smeets, U. F. Keyser, D. Krapf, M.-Y. Wu, N. H. Dekker, and C. Dekker, *Salt Dependence of Ion Transport and DNA Translocation through Solid-State Nanopores*, Nano Letters 6, 89 (2006).

[38] W. Guo, Y. Tian, and L. Jiang, *Asymmetric Ion Transport through Ion- Channel-Mimetic Solid-State Nanopores*, Accounts of Chemical Research 46, 2834 (2013).

[39] S. Ghosal, J. D. Sherwood, and H.-C. Chang, *Solid-state Nanopore Hydrodynamics and Transport*, Biomicrofluidics 13, 011301 (2019).

[40] N. Modi, M. Winterhalter, and U. Kleinekathoer, *Computational Modeling of Ion Transport through Nanopores*, Nanoscale 4, 6166 (2012).

[41] L. Restrepo-Pérez, S. John, A. Aksimentiev, C. Joo, and C. Dekker, *SDS assisted Protein Transport through Solid-state Nanopores*, Nanoscale 9, 11685 (2017).

[42] G. F. Schneider, S. W. Kowalczyk, V. E. Calado, G. Pandraud, H. W. Zandbergen, L. M. K. Vandersypen, and C. Dekker, *DNA Translocation through Graphene Nanopores*, Nano Letters 10, 3163 (2010).

[43] F. Traversi, C. Raillon, S. M. Benameur, K. Liu, S. Khlybov, M. Tosun, D. Krasnozhon, A. Kis, and A. Radenovic, *Detecting The Translocation of DNA Through A Nanopore Using Graphene Nanoribbons*, Nature Nanotechnology 8, 939 (2013).

[44] F. Al-Dirini, M. A. Mohammed, M. S. Hossain, F. M. Hossain, A. Nirmalathas, and E. Skafidas, *Tuneable Graphene Nanopores for Single Biomolecule Detection*, Nanoscale 8, 10066 (2016).

[45] C. R. Crick, J. Y. Y. Sze, M. Rosillo-Lopez, C. G. Salzmann, and J. B. Edel, *Selectively Sized Graphene-Based Nanopores for in Situ Single Molecule Sensing*, ACS Applied Materials & Interfaces 7, 18188 (2015).

[46] A. Szymczyk and P. Fievet, *Ion Transport Through Nanofiltration Membranes: The Steric, Electric And Dielectric Exclusion Model*, Desalination 200, 122 (2006), euromembrane 2006.

[47] W. Bowen and H. Mukhtar, *Characterisation And Prediction of Separation Performance of Nanofiltration Membranes*, Journal of Membrane Science 112, 263 (1996).

[48] A. E. Yaroshchuk, *Non-steric Mechanisms of Nanofiltration: Superposition of Donnan And Dielectric Exclusion*, Separation and Purification Technology 22-23, 143 (2001).

[49] H. Strathmann, Electrodialysis, *A Mature Technology With A Multitude of New Applications*, Desalination 264, 268 (2010)

[50] K. R. Williams and G. Burstein, *Low Temperature Fuel Cells: Interactions Between Catalysts And Engineering Design*, Catalysis Today 38, 401 (1997), fuel Cells and Catalysis.

[51] D. Konatham, J. Yu, T. A. Ho, and A. Striolo, *Simulation Insights for Graphene-Based Water Desalination Membranes*, Langmuir 29, 11884 (2013).

[52] D. Cohen-Tanugi, L.-C. Lin, J. C. Grossman, *Multilayer Nanoporous Graphene Membranes for Water Desalination*, Nano Letters 16, 1027 (2016).

[53] S. P. Surwade, S. N. Smirnov, I. V. Vlassiouk, R. R. Unocic, G. M. Veith, S. Dai, S. M. Mahurin, *Water Desalination Using Nanoporous Single-layer Graphene*, Nature Nanotechnology 10, 459 (2015).

[54] T. S. Plett, W. Cai, M. Le Thai, I. V. Vlassiouk, R. M. Penner, Z. S. Siwy,

Solid-State Ionic Diodes Demonstrated in Conical Nanopores, The Journal of Physical Chemistry C 121, 6170 (2017).

[55] S. Bukola, Y. Liang, C. Korzeniewski, J. Harris, and S. Creager, *Selective Proton/Deuteron Transport through Nafion—Graphene—Nafion Sandwich Structures at High Current Density*, Journal of the American Chemical Society 140, 1743 (2018).

[56] J. Gao, W. Guo, D. Feng, H. Wang, D. Zhao, and L. Jiang, *High-Performance Ionic Diode Membrane for Salinity Gradient Power Generation*, Journal of the American Chemical Society 136, 12265 (2014).

[57] T. Mouterde, A. Keerthi, A. R. Poggioli, S. A. Dar, A. Siria, A. K. Geim, L. Bocquet, and B. Radha, *Molecular Streaming and Its Voltage Control in Ångström-scale Channels*, Nature 567, 87 (2019).

[58] S. Hu, M. Lozada-Hidalgo, F. C. Wang, A. Mishchenko, F. Schedin, R. R. Nair, E. W. Hill, D. W. Boukhvalov, M. I. Katsnelson, R. A. Dryfe, I. V. Grigorieva, H. A. Wu, and A. K. Geim, *Proton Transport Through One- Atom-Thick Crystals*, Nature 516, 227 (2014).

[59] G.-R. Xu, J.-M. Xu, H.-C. Su, X.-Y. Liu, Lu-Li, H.-L. Zhao, H.-J. Feng, and R. Das, *Two-dimensional (2D) Nanoporous Membranes with Sub-nanopores in Reverse Osmosis Desalination: Latest Developments and Future Directions*, Desalination 451, 18 (2019), nanomaterials for Water Desalination: Recent Advances and Future Challenges.

[60] E. A. Jackson, M. A. Hillmyer, *Nanoporous Membranes Derived from Block Copolymers: From Drug Delivery to Water Filtration*, ACS Nano 4, 3548 (2010).

[61] M. I. Walker, K. Ubych, V. Saraswat, E. A. Chalklen, P. Braeuninger-Weimer, S. Caneva, R. S. Weatherup, S. Hofmann, and U. F. Keyser, *Extrinsic Cation Selectivity of 2D Membranes*, ACS Nano 11, 1340 (2017).

[62] L. Mogg, S. Zhang, G. P. Hao, K. Gopinadhan, D. Barry, B. L. Liu, H. M. Cheng, A. K. Geim, and M. Lozada-Hidalgo, *Perfect Proton Selectivity in Ion Transport Through Two-dimensional Crystals*, Nature Communications 10, 1 (2019).

[63] T. Jain, B. C. Rasera, R. J. S. Guerrero, M. S. Boutilier, S. C. O'Hern, J. C. Idrobo, and R. Karnik, *Heterogeneous Sub-continuum Ionic Transport in Statistically Isolated Graphene Nanopores*, Nature Nanotechnology 10, 1053 (2015)

[64] H. Yao, J. Zeng, P. Zhai, Z. Li, Y. Cheng, J. Liu, D. Mo, J. Duan, L. Wang, Y. Sun, and J. Liu, *Large Rectification Effect of Single Graphene Nanopore Supported by PET Membrane*, ACS Applied Materials and Interfaces 9, 11000 (2017).

[65] J. Feng, M. Graf, K. Liu, D. Ovchinnikov, D. Dumcenco, M. Heiranian, V. Nandigana, N. R. Aluru, A. Kis, and A. Radenovic, *Single-layer MoS2 Nanopores as Nanopower Generators*, Nature 536, 197 (2016).

[66] M. Caglar, I. Silkina, B. T. Brown, A. L. Thorneywork, O. J. Burton, V. Babenko, S. M. Gilbert, A. Zettl, S. Hofmann, and U. F. Keyser, *Tunable Anion-Selective Transport through Monolayer Graphene and Hexagonal Boron Nitride*, ACS Nano 14, 2729 (2020).

[67] Z. Li, Y. Liu, Y. Zhao, X. Zhang, L. Qian, L. Tian, J. Bai, W. Qi, H. Yao, B. Gao, J. Liu, W. Wu, and H. Qiu, *Selective Separation of Metal Ions via Monolayer Nanoporous Graphene with Carboxyl Groups*, Analytical Chemistry 88, 10002 (2016).

[68] Y. Li, G. Du, G. Mao, J. Guo, J. Zhao, R. Wu, and W. Liu, *Electrical Field Regulation of Ion Transport in Polyethylene Terephthalate Nanochannels*, ACS Applied Materials & Interfaces 11, 38055 (2019).

[69] R. M. Wyss, T. Tian, K. Yazda, H. G. Park, and C. J. Shih, *Macroscopic Salt Rejection through Electrostatically Gated Nanoporous Graphene*, Nano Letters 19, 6400 (2019).

[70] D. E. Goldman, Potential, *Impedence, And Rectification in Membranes*, Journal of General Physiology 27, 37 (1943).

[71] A. L. Hodgkin and R. D. Keynes, *The Potassium Permeability of A Giant Nerve Fibre*, The Journal of Physiology 128, 61 (1955).

[72] A. L. Hodgkin and A. F. Huxley, *A Quantitative Description of Membrane Current And Its Application to Conduction And Excitation in Nerve*, The Journal of Physiology 117, 500 (1952).

[73] J. Clay, *Determining K+ Channel Activation Curves from K+ Channel Currents Often Requires The Goldman-Hodgkin-Katz Equation*, Frontiers in Cellular Neuroscience 3, 20 (2009).

[74] R. C. Rollings, A. T. Kuan, and J. A. Golovchenko, *Ion Selectivity of Graphene Nanopores*, Nat. Commun. 7, 11408 (2016).

[75] P. M. G. van Deursen, Z. Tang, A. Winter, M. J. Mohn, U. Kaiser, A. A. Turchanin, and G. F. Schneider, *Selective Ion Sieving Through Arrays of Sub-nanometer Nanopores in Chemically Tunable 2D Carbon membranes,* Nanoscale 11, 20785 (2019).

[76] Cohen-Tanugi, D.; Grossman, J. C. *Water Desalination across Nanoporous Graphene*. Nano Lett. 2012, 12 (7), 3602-3608.

[77] J. J. Kasianowicz, E. Brandin, D. Branton, D. W. Deamer, *Characterization of individual polynucleotide molecules using a membrane channel*, Proc. Natl. Acad. Sci. USA 1996, 93, 13770.

[78] Dekker, C., *Solid-state Nanopores*, Nature Nanotechnology, vol 2, April 2007

[79] M. Wanunu, *Nanopores: A journey towards DNA sequencing*, Phys. Life Rev. 2012, 9, 125

[80] L. J. Steinbock, S. Krishnan, R. D. Bulushev, S. Borgeaud, M. Blokesch, L. Feletti, A. Radenovic, *Probing the size of proteins with glass nanopores*, Nanoscale 2014, 6, 14380.

[81] M. H. Lee, A. Kumar, K. B. Park, S. Y. Cho, H. M. Kim, M. C. Lim, Y. R. Kim, K. B. Kim, *A low-noise solid-state nanopore platform based on a highly insulating substrate*, Sci. Rep. 2014, 4, 7448.

[82] Suk, M.; Aluru, N. R. J., *Water Transport through Ultrathin Graphene*, Phys. Chem. Lett. 2010, 10 (1), 1590– 1594.

[83] Sint, K.; Wang, B.; Kral, P. J., *Selective Ion Passage through Functionalized Graphene Nanopores*, Am. Chem. Soc. 2008, 130 (49),16448–16449.

[84] D. W. Deamer, M. Akeson, *Nanopores and nucleic acids: prospects for ultrarapid sequencing,* Trends Biotechnol. 2000, 18, 147. B. M. Venkatesan, R. Bashir, Nat. Nanotechnol. 2011, 6, 615.

[85] B. M. Venkatesan, R. Bashir, *Nanopore sensors for nucleic acid analysis*, Nat. Nanotechnol. 2011, 6, 615.

[86] M. Tsutsui, M. Taniguchi, K. Yokota, T. Kawai, *Identifying single nucleotides by tunnelling current,* Nat. Nanotechnol. 2010, 5, 286;

[87] A. P. Ivanov, E. Instuli, C. M. McGilvery, G. Baldwin, D. W. McComb, T. Albrecht, J. B. Edel, *DNA tunneling detector embedded in a nanopore*, Nano Lett. 2011, 11, 279.

[88] G. A. Chansin, R. Mulero, J. Hong, M. J. Kim, A. J. DeMello, J. B. Edel, *Single-molecule spectroscopy using nanoporous membranes*, Nano Lett. 2007, 7, 2901

[89] M. Akeson, D. Branton, J. J. Kasianowicz, E. Brandin, D. W. Deamer, *Microsecond time-scale discrimination among polycytidylic acid, polyadenylic acid, and polyuridylic acid as homopolymers or as segments within single RNA molecules*. Biophys. J. 1999, 77, 3227.

[90] A. Darvish, G. Goyal, R. Aneja, R. V. Sundaram, K. Lee, C. W. Ahn, K. B. Kim, P. M. Vlahovska, M. J. Kim, *Nanoparticle mechanics: deformation detection via nanopore resistive pulse sensing*. Nanoscale 2016, 8, 14420

[91] G. Ando, C. Hyun, J. Li, T., *Mitsui, Directly Observing the Motion of DNA Molecules near Solid-State Nanopores*, ACS Nano 2012, 6, 10090.

[92] W. H. Pitchford, H. J. Kim, A. P. Ivanov, H. M. Kim, J. S. Yu, R. J. Leatherbarrow, T. Albrecht, K. B. Kim, J. B. Edel, *Synchronized Optical and Electronic Detection of Biomolecules Using a Low Noise Nanopore Platform*, ACS Nano 2015, 9, 1740.

[93] S. Banerjee, J. Wilson, J. Shim, M. Shankla, E. A. Corbin, A. Aksimentiev, R. Bashir, *Slowing DNA Transport Using Graphene-DNA Interaction*, Adv. Funct. Mater. 2015, 25, 936

[94] Garaj, S. W. Hubbard, A. Reina, J. Kong, D. Branton, J. A. Golovchenko, *Graphene as a subnanometre trans-electrode membrane*, Nature 467, 190-193 (2010).

[95] Merchant, C. A., Healy, K., Wanunu, M., Ray, V., Peterman, N., Bartel, J., Fischbein, M.D., Venta, K., Luo, Z., Johnson, A.T.C. Drndić, M., *DNA Translocation through Graphene Nanopores*, Nano Lett. 10, 2915-2921 (2010).

[96] Koenig, S. P.; Wang, L.; Pellegrino, J.; Bunch, J. S., *Selective Molecular Sieving through Porous Graphene*. Nat. Nanotechnol. 2012, 7 (11), 728-732.

[97] D. Cohen-Tanugi, J.C. Grossman, *Mechanical strength of nanoporous graphene as a desalination membrane,* Nano Lett. 14 (2014) 6171e6178.

[98] J. Lee, A.K. Roy, J.L. Wohlwenk, V. Varshney, J.B. Ferguson, W.C. Mirchel, *Scaling law for energy bandgap and effective electron mass in graphene nano mesh*, Appl. Phys. Lett. 102 (2013) 203107.

[99] F. Ouyang, S. Peng, Z. Liu, *Bandgap opening in graphene antidot lattices: the missing half*, ACS Nano 5 (2011) 4023e4030.

[100] https://www.ks.uiuc.edu/Gallery/Science-New/

[101] L. Hu, S. Wyant, A.R. Muniz, A. Ramasubramaniam, D. Maroudas, *Mechanical behavior and fracture of graphene nanomeshes*, J. Appl. Phys. 117 (2015), 024302.

[102] C. Carpenter, A.M. Christmann, L. Hu, L. Fampiou, A.R. Muniz, A. Ramasubramaniam, *Elastic properties of graphene nanomeshes*, Appl. Phys. Lett. 104 (2014) 141911.

[103] Y. Liu, X. Chen, *Mechanical properties of nanoporous graphene membranes*, J. Appl. Phys. 115 (2014), 034303.

[104] A. Meller, L. Nivon, E. Brandin, J. Golovchenko, D. Branton, *Rapid nanopore discrimination between single polynucleotide molecules*, Proc. Natl. Acad. Sci. USA 2000, 97, 1079

[105] M. Wanunu, T. Dadosh, V. Ray, J. Jin, L. McReynolds, M. Drndic, *Rapid electronic detection of probe-specific microRNAs using thin nanopore sensors*, Nat. Nanotechnol. 2010, 5, 807.

[106] Huang, H., Wei, D., Sun, J., Wong, S.L., Feng, Y.P., Castro Neto, A.H., Wee, A.T.S, *Spatially Resolved Electronic Structures of Atomically Precise Armchair Graphene Nanoribbons*, Sci. Rep. 2, 983, (2012)

[107] Sinitskii, A., *A recipe for nanoporous graphene*, Science 360 (6385), 154-155, 2018

[108] Bieri, M., Treier, M., Cai, J., Aït-Mansour, K., Ruffieux, P., Gröning, O., Gröning, P., Kastler, M., Rieger, R., Feng, X., Müllen, K., Fasel, R., *Porous graphenes: two-dimensional polymer synthesis with atomic precision,* Chem. Commun., 2009, 6919-6921

[109] Gutzler, R., Walch, H., Eder, G., Kloft, S., Hecklab, W.M., Lackinger, M., *Surface mediated synthesis of 2D covalent organic frameworks: 1,3,5-tris(4-bromophenyl)benzene on graphite(001), Cu(111), and Ag(110)*, Chem. Commun., 2009, 4456-4458

[110] Teeter, J.D., Costa, P.S., Zahl, P., Vo, T.H., Shekhirev, M., Xu, W., Zeng, X.C., Enders, A., Sinitskii, A., *Dense monolayer films of atomically precise graphene nanoribbons on metallic substrates enabled by direct contact transfer of molecular precursors*, Nanoscale, 9, 18835, (2017)

[111] Llinas, J.P., Fairbrother, A., Barin, G.B., Shi, W., Lee , K., Wu, S., Choi, B.Y., Braganza, R., Lear, J., Kau, N., Choi, W. Chen, C., Pedramrazi, Z., Dumslaf, T., Narita, A., Feng, X., Müllen, K., Fischer, F., Zettl, A., Ruffieux, P., Yablonovitch, E., Crommie, M., Fasel, R., Bokor, J., *Short-channel field-effect transistors with 9-atom and 13-atom wide graphene nanoribbons,* Nat. Commun. 8, 633 (2017)

[112] J. Bai, X. Zhong, S. Jiang, Y. Huang, X. Duan, *Graphene nanomesh*, Nat. Nanotechnol. 5 (2010)

[113] Yuan, W., Chen, J., Shi, G., *Nanoporous graphene materials*, Mater. Today 17 (2014)